FEDERAL LOAN GUARANTEES FOR ENERGY PROJECTS

ELEMENTS AND CONSIDERATIONS

ENERGY POLICIES, POLITICS AND PRICES

Additional books in this series can be found on Nova's website under the Series tab.

Additional E-books in this series can be found on Nova's website under the E-book tab.

GOVERNMENT PROCEDURES AND OPERATIONS

Additional books in this series can be found on Nova's website under the Series tab.

Additional E-books in this series can be found on Nova's website under the E-book tab.

ENERGY POLICIES, POLITICS AND PRICES

FEDERAL LOAN GUARANTEES FOR ENERGY PROJECTS

ELEMENTS AND CONSIDERATIONS

RONAN SAMPSON

AND

CAMERON LANGE

EDITORS

New York

NOTICE TO THE READER

LIBRARY OF CONGRESS CATALOGING-IN-PUBLICATION DATA

ISBN: 978-1-62417-116-1

Published by Nova Science Publishers, Inc. † New York

CONTENTS

PREFACE

The federal government has a number of policy tools available to encourage the development and deployment of innovative clean energy technologies. Some of these policy tools include: clean energy mandates, carbon taxes, carbon cap and trade, environmental regulations, loan guarantees, grants, and tax expenditures. In 2005 and 2009, Congress passed legislation that provided loan guarantee authority to the Department of Energy (DOE) for innovative clean energy technologies and clean energy projects. In 2011, the high-profile bankruptcy, and subsequent loan default, of Solyndra resulted in a congressional investigation and subjected DOE's loan guarantee program to a high degree of scrutiny. This book provides analysis of goals for and concerns about the use of loan guarantees as a mechanism to support the deployment of innovative clean energy technologies.

Chapter 1 – Government guaranteed debt is a financial tool that has been used to support a number of federal policy objectives: home ownership, higher education, and small business development, among others. Loan guarantees for new energy technologies date back to the mid-1970s, when rapidly rising energy prices motivated the development of alternative, and renewable, sources of energy. Recently, the Energy Policy Act of 2005 created a loan guarantee program for innovative clean energy technologies (nuclear, clean coal, renewables) commonly known as Section 1703. The American Recovery and Reinvestment Act of 2009 created Section 1705, a temporary loan guarantee program focused on deployment of renewable energy technologies and projects.

Loan guarantee authority for the Department of Energy Loan Programs Office (LPO) Section 1705 program ended on September 30, 2011, prior to which approximately $16.15 billion of loans were guaranteed for a variety of clean energy projects. In August 2011, the high-profile bankruptcy of

Solyndra, the first company to receive a Section 1705 loan guarantee, resulted in a congressional investigation and increased scrutiny of the DOE Loan Guarantee Program. As a result, Congress may decide to evaluate the use of loan guarantees as a mechanism for supporting the development and deployment of clean energy technologies. This report analyzes goals and concerns associated with innovative clean energy loan guarantees.

Fundamentally, loan guarantees can provide access to low-cost capital for projects that might be considered high risk by the commercial banking and investment community. There are many goals for using loan guarantees to support innovative energy technology commercialization and deployment. Commercializing new technologies that may increase the performance and reduce the cost of clean energy generation is one objective. Also, the potential global market for clean energy technologies and systems is substantial (trillions of dollars over the next 25 years by some estimates) and loan guarantees could help position U.S. manufacturers to supply product for this growing market. Loan guarantees may also result in near- and long-term job creation as well as contribute toward reducing emissions of various pollutants.

The high-risk nature of clean energy projects, however, raises some concerns about the use of loan guarantees as a mechanism to encourage the deployment of new technologies. First, loan repayment demands cash flow from development stage companies at a time when they may already have high cash flow requirements, so loan repayment obligations could actually increase the risk of default for certain projects. Second, at a project level, the government's potential return is not commensurate with the risk being assumed. Third, loan guarantees for clean energy technologies are essentially long-term commitments in a dynamic and evolving marketplace. As a result, technologies supported today could be obsolete in less than a decade, thereby increasing the risk of loan default. Finally, federally managed loan guarantee programs may be subject to certain pressures that could result in less-than-optimal decision making.

Should Congress decide to continue the use of government financial tools as a clean energy technology deployment support mechanism, it may wish to consider various policy options for future initiatives. Some policy options could include (1) using grants or tax expenditures instead of loan guarantees; (2) taking equity positions in new technologies and projects through a new government-backed venture-capital-like organization; (3) authorizing the use of flexible management tools such as stock warrants, portfolio management, and convertible equity; and (4) creating a dedicated clean energy financial support authority to manage federal clean energy deployment investments. Each of these policy options is explored and discussed in this report.

Chapter 2 – Since Solyndra, a solar system manufacturing company that received a $535 million loan guarantee from the Department of Energy (DOE), filed for bankruptcy in September of 2011 there has been much congressional interest in better understanding the characteristics of renewable energy projects, specifically solar projects, that have received DOE loan guarantees. The objective of this report is to provide Congress with insight regarding solar projects supported by DOE's loan guarantee program, the risk characteristics of these projects, and how other DOE loan guarantee projects are either similar to or different from the Solyndra solar manufacturing project.

Chapter 3 – The Department of Energy's (DOE) Loan Guarantee Program (LGP) was created by section 1703 of the Energy Policy Act of 2005 to guarantee loans for innovative energy projects. Currently, DOE is authorized to make up to $34 billion in section 1703 loan guarantees. In February 2009, the American Recovery and Reinvestment Act added section 1705, making certain commercial technologies that could start construction by September 30, 2011, eligible for loan guarantees. It provided $6 billion in appropriations that were later reduced by transfer and rescission to $2.5 billion. The funds could cover DOE's costs for an estimated $18 billion in additional loan guarantees. GAO has an ongoing mandate to review the program's implementation. Because of concerns raised in prior work, GAO assessed (1) the status of the applications to the LGP and (2) for loans that the LGP has committed to, or made, the extent to which the program has adhered to its process for reviewing applications. GAO analyzed relevant legislation, regulations, and guidance; prior audits; and LGP data, documents, and applications. GAO also interviewed DOE officials and private lenders with experience in energy project lending.

In: Federal Loan Guarantees for Energy Projects   ISBN: 978-1-62417-116-1
Editors: R. Sampson and C. Lange   © 2013 Nova Science Publishers, Inc.

Chapter 1

LOAN GUARANTEES FOR CLEAN ENERGY TECHNOLOGIES: GOALS, CONCERNS, AND POLICY OPTIONS[*]

Phillip Brown

SUMMARY

Government guaranteed debt is a financial tool that has been used to support a number of federal policy objectives: home ownership, higher education, and small business development, among others. Loan guarantees for new energy technologies date back to the mid-1970s, when rapidly rising energy prices motivated the development of alternative, and renewable, sources of energy. Recently, the Energy Policy Act of 2005 created a loan guarantee program for innovative clean energy technologies (nuclear, clean coal, renewables) commonly known as Section 1703. The American Recovery and Reinvestment Act of 2009 created Section 1705, a temporary loan guarantee program focused on deployment of renewable energy technologies and projects.

Loan guarantee authority for the Department of Energy Loan Programs Office (LPO) Section 1705 program ended on September 30, 2011, prior to which approximately $16.15 billion of loans were guaranteed for a variety of clean energy projects. In August 2011, the

---

[*] This is an edited, reformatted and augmented version of a Congressional Research Service publication, CRS Report for Congress R42152, from www.crs.gov, prepared for Members and Committees of Congress, dated January 17, 2012.

high-profile bankruptcy of Solyndra, the first company to receive a Section 1705 loan guarantee, resulted in a congressional investigation and increased scrutiny of the DOE Loan Guarantee Program. As a result, Congress may decide to evaluate the use of loan guarantees as a mechanism for supporting the development and deployment of clean energy technologies. This report analyzes goals and concerns associated with innovative clean energy loan guarantees.

Fundamentally, loan guarantees can provide access to low-cost capital for projects that might be considered high risk by the commercial banking and investment community. There are many goals for using loan guarantees to support innovative energy technology commercialization and deployment. Commercializing new technologies that may increase the performance and reduce the cost of clean energy generation is one objective. Also, the potential global market for clean energy technologies and systems is substantial (trillions of dollars over the next 25 years by some estimates) and loan guarantees could help position U.S. manufacturers to supply product for this growing market. Loan guarantees may also result in near- and long-term job creation as well as contribute toward reducing emissions of various pollutants.

The high-risk nature of clean energy projects, however, raises some concerns about the use of loan guarantees as a mechanism to encourage the deployment of new technologies. First, loan repayment demands cash flow from development stage companies at a time when they may already have high cash flow requirements, so loan repayment obligations could actually increase the risk of default for certain projects. Second, at a project level, the government's potential return is not commensurate with the risk being assumed. Third, loan guarantees for clean energy technologies are essentially long-term commitments in a dynamic and evolving marketplace. As a result, technologies supported today could be obsolete in less than a decade, thereby increasing the risk of loan default. Finally, federally managed loan guarantee programs may be subject to certain pressures that could result in less-than-optimal decision making.

Should Congress decide to continue the use of government financial tools as a clean energy technology deployment support mechanism, it may wish to consider various policy options for future initiatives. Some policy options could include (1) using grants or tax expenditures instead of loan guarantees; (2) taking equity positions in new technologies and projects through a new government-backed venture-capital-like organization; (3) authorizing the use of flexible management tools such as stock warrants, portfolio management, and convertible equity; and (4) creating a dedicated clean energy financial support authority to manage federal clean energy deployment investments. Each of these policy options is explored and discussed in this report.

INTRODUCTION

The federal government has a number of policy tools available to encourage the development and deployment of innovative clean energy technologies (see text box below). Some of these policy tools include (1) clean energy mandates, (2) carbon taxes, (3) carbon cap and trade, (4) environmental regulations, (5) loan guarantees, (6) grants, and (7) tax expenditures. In 2005, Congress passed legislation that provided loan guarantee authority to the Department of Energy (DOE) for innovative clean energy technologies. In 2009, Congress passed legislation that modified DOE's loan guarantee authority and created a temporary loan guarantee program for the deployment of clean energy technologies and the development of clean energy projects. In 2011, the high-profile bankruptcy, and subsequent loan default, of Solyndra resulted in a congressional investigation and subjected DOE's loan guarantee program to a high degree of scrutiny.[1]

This report provides analysis of goals for and concerns about the use of loan guarantees as a mechanism to support the deployment of innovative clean energy technologies.

A discussion of several policy options for Congress to consider is also provided, should Congress decide to debate the future of clean energy loan guarantee programs.

What Are "Innovative Clean Energy Technologies?"

Many different types of energy technologies could be considered "innovative" and "clean." The innovative aspect of new technologies typically refers to a new approach or method that can either increase the performance of energy generation technologies and/or reduce the cost of producing useable forms of energy, such as electricity or fuels. Typically, innovative technologies have been demonstrated to some degree but are not yet available in the commercial marketplace.

The clean aspect of new energy technologies usually refers to the ability of a technology to reduce or eliminate the amount of emissions (e.g., carbon) per unit of energy produced. Renewable energy technologies such as solar, wind, goethermal, biomass, biofuels, and others are almost always categorized as "clean." However, advanced nuclear and clean coal technologies might also be considered "clean" based on their ability to reduce carbon emissions.

BACKGROUND AND HISTORY OF FEDERAL LOAN GUARANTEES

A loan guarantee might be defined as "a loan or security on which the federal government has removed or reduced a lender's risk by pledging to repay principal and interest in case of default by the borrower."[2] Historically, loan guarantees have been used as a policy tool for many different purposes, including home ownership, university education, small business growth, international development, and others.

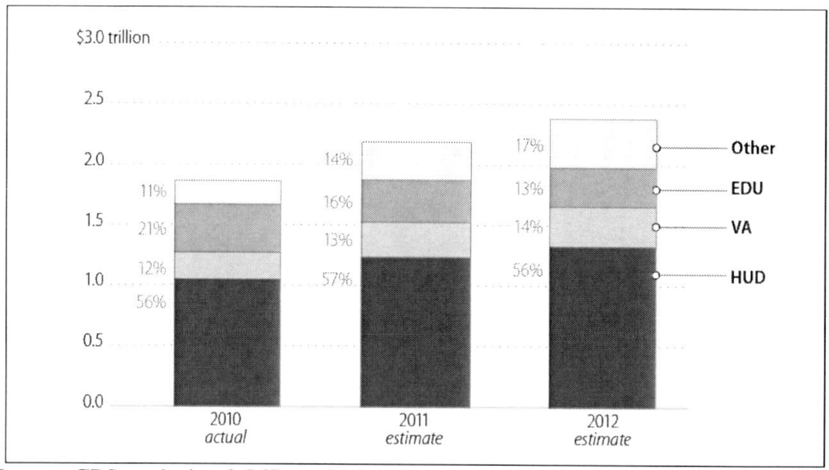

Source: CRS analysis of Office of Management and Budget FY 2012 budget, Table 23-12 "Guaranteed Loan Transactions of the Federal Government," available at http://www.whitehouse.gov/sites/default/files/omb/budget/ fy2012/assets/23_12.pdf.

Notes: According to OMB Circular A-11 (Revised November 2011), federal loan guarantees include the full face value of outstanding loan guarantees. Full face value includes both the guaranteed and non-guaranteed portion of loan principal outstanding. Therefore, the actual amount of federal government liability may not be accurately reflected in this figure. Also, "Primary Guaranteed Loans" are calculated by summing the face value of all federal guaranteed loans and then adjusting downward to account for secondary loan guarantees.

EDU = Department of Education

VA = Department of Veterans Affairs

HUD = Department of Housing and Urban Development

Figure 1. U.S. Federal Government Primary Guaranteed Loans Outstanding.

Today, 14 federal government agencies manage approximately 68 loan guarantee accounts that include approximately $1.9 trillion of primary guaranteed loans outstanding in 2010 (see Figure 1).[3] Primary guaranteed loan amounts include the total face value of the loans and not just the federally guaranteed portion of those loans.[4]

The first large-scale use of federal loan guarantees occurred during the 1930s Great Depression, when loan guarantees were used as a mechanism to assist families with purchasing homes. Home purchase loan guarantees are designed to be actuarially sound by charging borrowers insurance fees, which are pooled and used to pay for program operating costs and probable losses associated with loan defaults. Loan guarantees have also been used for higher risk borrowers such as students or low-income families. These borrowers might be considered higher risk because of a greater likelihood of default or inadequate collateral to support a loan. As a result, the government bears a portion of the default risk when lending to these types of borrowers; therefore these loans generally include some degree of government subsidy.[5]

Concerns about budgetary reporting of loan guarantees resulted in the Federal Credit Reform Act of 1990 (FCRA), which was included in the Omnibus Budget Reconciliation Act of 1990 (P.L. 101-508). Prior to the enactment of FCRA, fiscal year cash flow accounting was used to report the budgetary costs of loan guarantees, and this approach did not accurately take into account the expected losses associated with loan guarantee programs. Therefore, the total cost of long-term loan guarantees was not adequately accounted for, and reported, in the short-term congressional budget window. FCRA mandated an accrual accounting approach for budget reporting and required that budgetary costs of loan guarantees be reported as the net present value of subsidy costs associated with long-term loan guarantees.[6] The Office of Management and Budget provides guidance for calculating the credit subsidy cost to agencies that administer loan guarantee programs.[7]

Loan guarantees have also been used to finance relatively large (from $10 million to over $1 billion) energy and infrastructure projects. Programs for such projects typically consist of a small number of projects with large capital requirements. As a result, it is difficult for loan guarantee programs for these types of projects to be actuarially sound because there is not a large enough project pool to spread the risk. While federal credit guidelines require credit subsidy costs (much like a loan loss reserve) for loan guarantee projects be collected, these costs are typically paid for through federally appropriated funds.

Congress has two primary mechanisms for controlling federal loan guarantee programs. First, Congress can appropriate funds to pay for credit subsidy costs, and this approach can limit the amount of federally supported loan guarantees once the credit subsidy appropriation has been exhausted. Second, Congress can stipulate volume limits for loan guarantee programs. For example, Congress could limit the total value of loans supported by a certain program to $20 billion.

LOAN GUARANTEES FOR INNOVATIVE CLEAN ENERGY TECHNOLOGIES

Federal loan guarantee authorizations for demonstrating alternative energy technologies date back to the 1970s, when the Geothermal Energy Research, Development, and Demonstration Act of 1974 (P.L. 93-410) authorized loan guarantees for geothermal demonstration facilities.[8] The Department of Energy Act—Civilian Applications (P.L. 95-238), which became law in 1978, authorized the Secretary of Energy to guarantee loans for alternative fuel demonstration facilities.[9] In response to an energy price shock in 1979, Congress passed the Energy Security Act of 1980 (P.L. 96-294) that authorized $20 billion to create a domestic synthetic fuels industry through the use of loans, loan guarantees, price guarantees, joint ventures, and fuel purchase agreements. To execute this endeavor, the law established the quasi-public U.S. Synthetic Fuels Corporation (SFC), although the Department of Energy was authorized to fund projects prior to the official start-up of SFC.[10] Five projects were supported by the SFC, only one of which utilized a loan guarantee. The Great Plains coal gasification project (located in Beulah, ND), which converts lignite coal into pipeline-quality methane (the primary component of natural gas), received a $2.02 billion federal loan guarantee (approximately $1.5 billion of the loan guarantee was actually used) to construct the plant.[11] Due to energy price declines in the mid-1980s, along with a denied request to restructure debt and institute price support mechanisms, the Great Plains project was not able to meet debt service requirements and subsequently defaulted on its loan obligations in August 1985.[12] After paying off the defaulted loan, DOE proceeded to sell the Great Plains facility, which was purchased by Basin Electric for an initial price of $85 million.[13] Basin Electric assumed ownership of the plant on October 31,

1988.[14] Today, the Great Plains facility is operated by the Dakota Gasification Company, a subsidiary of Basin Electric.

The Energy Security Act of 1980 (P.L. 96-294) also resulted in the creation of the Office of Alcohol Fuels (OAF) within the Department of Energy. OAF was given the authority to guarantee loans for alcohol fuel projects and eventually guaranteed loans totaling approximately $265 million for three alcohol fuel projects. Of the three projects that received DOE loan guarantees, one had to refinance its loan, one experienced technology performance complications, and one ceased operations.[15]

Most recently, loan guarantees have been used as a mechanism to encourage development and deployment of innovative clean energy technologies.[16]

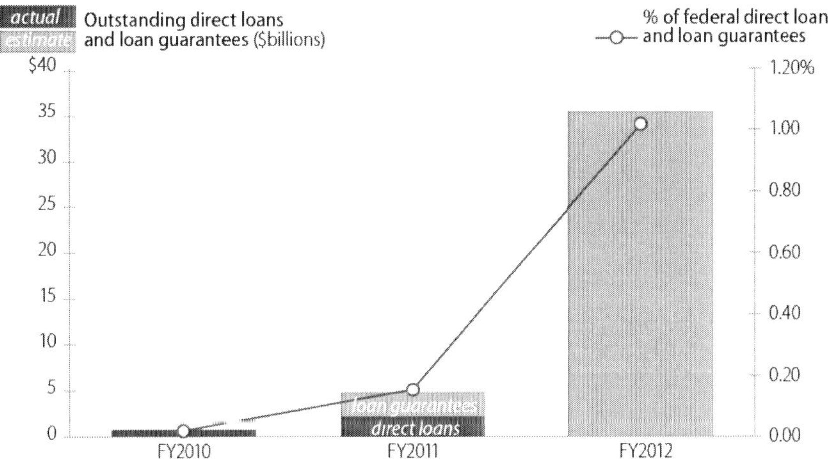

Source: CRS analysis of Office of Management and Budget FY 2012 budget, Table 23-11 "Direct Loan Transactions of the Federal Government," and Table 23-12 "Guaranteed Loan Transactions of the Federal Government," 2011 Direct Loan actual numbers were sourced from the Federal Financing Bank October 2011 activity report, available at http://www.treasury.gov/ffb/press_releases/2011/11-2011.shtml.

Notes: OMB data used for this figure is from the FY2012 budget, which was released in February 2011. Updated estimates for 2011 loan guarantees and 2012 loans and loan guarantees will be available in February 2012. Actuals and new estimates may be different than information provided in this figure.

Figure 2. Federal Credit Programs for Innovative Clean Energy Technologies (Direct loans and loan guarantees).

The Energy Policy Act of 2005 and the American Recovery and Reinvestment Act of 2009 resulted in the creation of DOE's Loan Programs Office (LPO), which was chartered to administer clean energy loan guarantee initiatives.[17] Loan guarantees for innovative clean energy technologies constitute a small but growing portion of federal direct loans and loan guarantees (see *Figure 2*).[18]

Energy Policy Act of 2005

The Energy Policy Act of 2005 (EPACT 2005; P.L. 109-58), enacted on August 8, 2005, established loan guarantee programs for multiple energy technologies. EPACT 2005 enabled loan guarantees to be used in support of projects for (1) commercial byproducts from municipal solid waste and cellulosic biomass, (2) sugar ethanol, (3) integrated coal/renewable energy systems, (4) coal gasification, (5) petroleum coke gasification, and (6) electricity production on Indian lands, among others. Title XVII of EPACT 2005 created a new loan guarantee program for these innovative energy technologies.

Title XVII—Incentives for Innovative Technologies

Title XVII of EPACT 2005 authorized the Department of Energy to provide loan guarantees for eligible innovative technologies that are not yet commercially available.[19] Projects eligible for federal loan guarantees, per Section 1703 of Title XVII, include a variety of technologies such as renewable energy systems, advanced fossil energy technologies, advanced nuclear technologies, and many others.

Title XVII also stipulates that no loan guarantees shall be made to projects unless the cost of the project is paid for by either (1) appropriated funds, or (2) the borrower. The definition of "cost" is based on that provided in Section 502(5)(C) of the Federal Credit Reform Act of 1990.[20] No funds were initially appropriated to pay for costs of loan guarantees provided under Title XVII, therefore borrowers were expected to pay for all loan guarantee costs.

American Recovery and Reinvestment Act of 2009

The American Recovery and Reinvestment Act of 2009 (ARRA 2009; P.L. 111-5) modified Title XVII of EPACT 2005 in two ways. First, ARRA established Section 1705, a temporary loan guarantee program for deployment of renewable energy and electricity transmission systems. Section 1705 loan guarantee authority ended on September 30, 2011. Second, ARRA 2009 included a $6 billion appropriation to pay for subsidy costs associated with projects authorized under the temporary Section 1705 program. This amount was reduced to $2.435 billion after rescissions and transfers.[21]

DOE's Loan Programs Office

To execute and administer federal credit programs for innovative energy technologies, the Department of Energy created its Loan Programs Office (LPO).[22] LPO administers three loan programs:

1. *Section 1703:* loan guarantees for innovative clean energy technologies with high degrees of technology risk.
2. *Section 1705:* loan guarantees for certain renewable energy systems, electric power transmission, and innovative biofuel projects that may have varying degrees (high or low) of technology risk.
3. *Advanced Technology Vehicle Manufacturing (ATVM):* direct loans to support advanced technology vehicles and associated components.[23]

As of December 2011, all finalized loan guarantee commitments have been for 28 projects within LPO's Section 1705 program, which equal approximately $16.15 billion of federal loan guarantee commitments. LPO's Section 1703 program has issued conditional loan guarantee commitments to four projects with a total loan guarantee value of approximately $10.6 billion.[24] *Figure 3* illustrates how Section 1705 loan guarantee commitments were distributed by technology types.

New Technology Deployment vs. Project Finance

Two general types of financing activities can be supported by loan guarantee programs for innovative clean energy technologies. The first type of finance activity is categorized as "new technology deployment." New technology deployment, for the purpose of this report, might include projects such as building a new manufacturing facility for a new energy technology (solar modules, wind turbines).

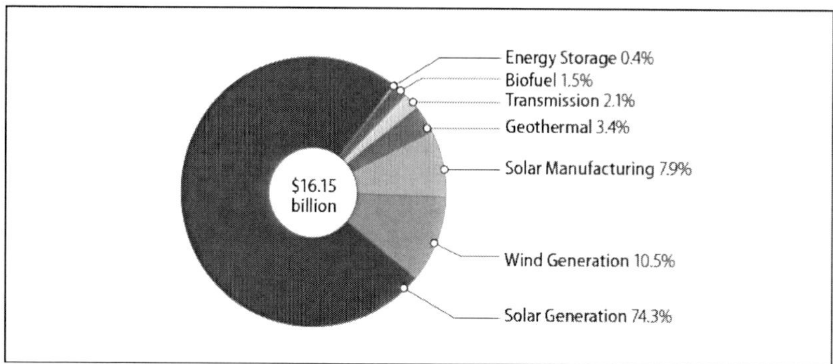

Source: CRS analysis of DOE Section 1705 loan guarantee recipients.
Notes: Numbers may not equal 100% due to rounding.

Figure 3. Section 1705 Loan Guarantees By Technology Category.

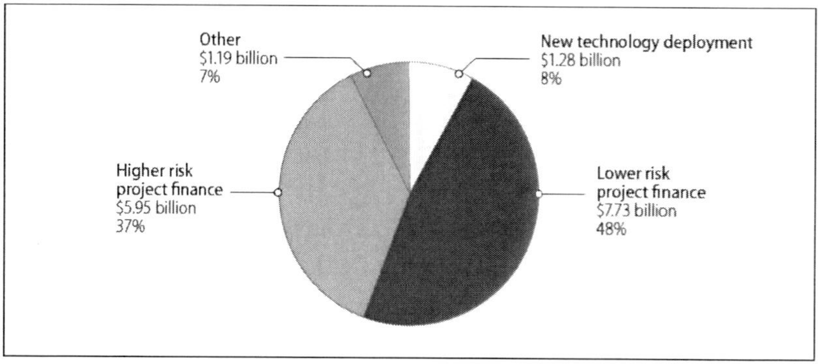

Source: CRS analysis of projects that received loan guarantees from the Department of Energy Loan Programs Office, Section 1705 program.
Notes: Projects classified as "New technology deployment" include loan guarantees that supported manufacturing of new energy technologies. The "Lower risk project finance" category includes electricity or fuel production projects that use commercially available technologies (most of this category consists of solar photovoltaic and wind projects). Projects classified as "Higher risk project finance" include electricity generation and fuel production projects that use technologies that might be considered less commercial (most projects in this category use some type of solar thermal technology).

Figure 4. DOE Section 1705 Loan Guarantees (New technology deployment and project finance).

Project finance includes projects that will use commercial, or near-commercial, technologies to generate electricity that will be purchased by a third party. Of the two financing types, new technology deployment projects are generally considered higher risk due to external technology and market dynamics that can significantly impact the financial performance of such projects. Project finance projects typically have lower risk profiles due to their ability to utilize contractual mechanisms (power purchase agreements, technology performance guarantees) as a means to minimize financial risk.[25] However, all project finance projects are not equal and the financial risk profile for these projects could be impacted by technology type, possible construction delays, and/or operations and maintenance characteristics.[26] *Figure 4* provides an assessment of the types of projects supported by DOE's Section 1705 loan guarantee program.

GOALS FOR CLEAN ENERGY LOAN GUARANTEES

One primary objective for providing federal loan guarantees for clean energy technologies and projects is to provide access to low cost financial capital that might not otherwise be available due to certain technology and market risks. Access to such capital may result in achieving certain policy objectives, assuming loan guarantee projects are successful and realize anticipated outcomes. Using loan guarantees as a mechanism for supporting U.S. clean energy technology deployment, project development, and system manufacturing can help meet various policy goals. Some of those goals are discussed below.

Commercialization of Innovative Technologies

Renewable energy technologies typically follow a common commercialization development path. Development of new technologies generally consists of the following stages: (1) feasibility analysis, (2) research and development, (3) system demonstration, (4) system scale-up and operation, and (5) commercial deployment (see *Figure 5*). Various federal government incentives can be used to support every stage of technology commercialization. However, the focus of this report is on system scale-up and commercial deployment due to the high-risk nature of these activities and the large amounts of capital required. Typically as technologies move through the development life cycle, the cost to complete each subsequent development stage increases, and in some cases the cost increases can be substantial. System scale-up and operation, and commercial deployment, are usually the most costly development stages. Financing these development activities can sometimes be difficult because the capital requirements are large and the risks (technology performance, market dynamics) are usually high. Some people refer to this situation as the "valley of death" or the "chasm" that all new technologies might encounter as they move from demonstration to commercial deployment. Formulating and executing a plan to realize commercial deployment is a challenge in itself.[27] Financing that plan can further complicate new technology commercialization. By providing a source of low-cost capital for these development stages, loan guarantees could support the commercialization of new and innovative renewable energy technologies.

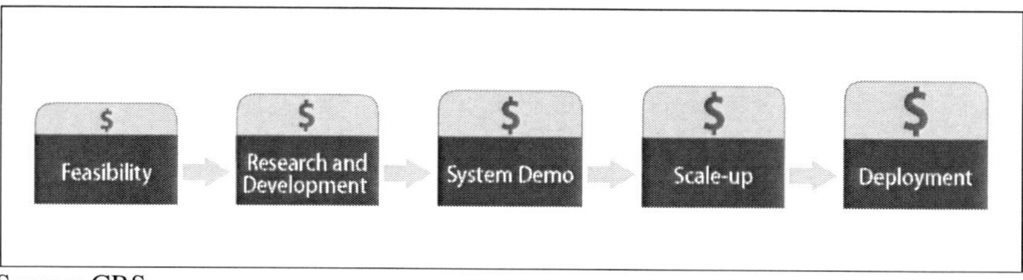

Source: CRS.

Figure 5. Notional Technology Development and Commercialization Lifecycle.

Positioning U.S. Manufacturing for an Emerging Global Market

Global renewable energy use is expected to grow. For example, the International Energy Agency (IEA) estimates, under one scenario, that electricity generation from renewable energy will grow from 3% of global electricity in 2009 to approximately 15% by 2035 (see *Figure 6*).[28] In order to realize these projections, IEA estimates that approximately $6 trillion of investment in renewable electricity generation will be needed between now and 2035.[29] As global renewable electricity markets expand, many countries may look to position themselves as leading manufacturers of renewable electricity generation systems and technologies. Loan guarantees for renewable electricity technology manufacturers could provide a source of low cost financial capital that might incentivize build-out of U.S. renewable energy manufacturing capacity. This capacity build-out could potentially result in economies of scale and make U.S. manufacturing cost competitive. If global markets expand as projected, U.S. manufacturers could be positioned to manufacture and export renewable energy technologies and systems for the global marketplace.

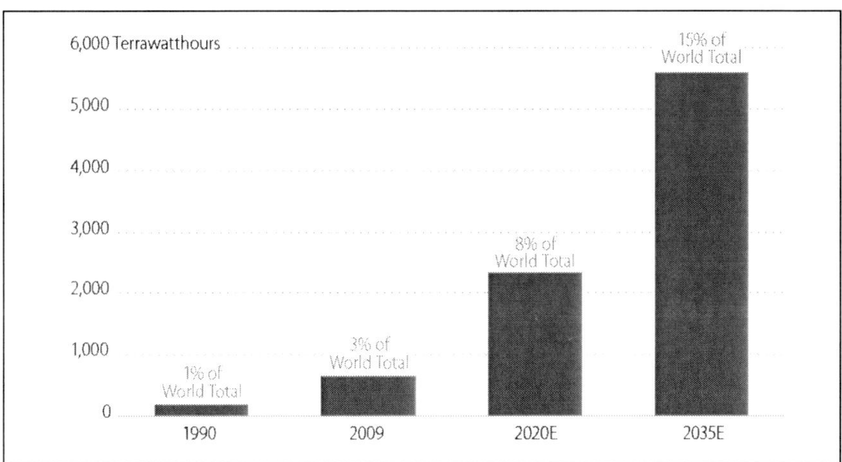

Source: International Energy Agency, World Energy Outlook 2011.

Notes: IEA's World Energy Outlook includes analysis for three different world energy scenarios: (1) Current Policies, (2) New Policies, and (3) the 450 scenario (referring to a limit on atmospheric carbon dioxide concentrations of 450 parts per million). This figure reflects IEA renewable electricity projections under the "New Policies" scenario.

Figure 6. Global Electricity Generation from Non-hydro Renewables.

Job Creation

Loan guarantees might result in job creation as a result of building and operating projects that utilize loan guarantee finance mechanisms and possibly through the expansion of new industries that establish a competitive position in the global marketplace. According to DOE's Loan Programs Office, jobs related to fully committed Section 1705 loan guarantees include approximately 14,300 construction jobs and 2,400 permanent jobs.[30] Construction jobs are typically temporary in nature, while permanent jobs are functions required to operate projects over their respective lifetimes. Additional job creation might occur if projects supported by loan guarantees are successful and realize their commercial deployment goals and objectives. The number of jobs that might ultimately result from loan guarantee projects that become globally competitive is difficult to estimate at this time due to unknown market, technology, and policy variables that will likely determine future renewable energy market growth.

Reducing Greenhouse Gas Emissions

Deployment of clean energy technologies and projects could potentially support greenhouse gas reduction goals, for example, by increasing the total amount of electricity generation from low carbon sources. Emission reductions that are directly associated with projects supported by DOE loan guarantees will likely be modest due to the massive scale of the U.S. energy industry.

However, larger indirect emission reductions may be achieved as a result of future deployment of clean energy projects, should loan guarantee projects achieve their objectives.

Supply Chain Build-Out

On its website, DOE's Loan Programs Office emphasizes that loan guarantees provided by LPO are supporting some of the largest solar photovoltaic, solar thermal, and wind electricity generation projects in the world.[31] Developing and constructing these large-scale projects may require domestic supply chains to support deployment of certain technologies. As a result, loan guarantees may support the build-out of a U.S. supply chain for clean energy technology, system, component, and logistics companies.[32] This build-out may help position these companies for global clean energy opportunities.

CONCERNS ABOUT LOAN GUARANTEES FOR INNOVATIVE ENERGY TECHNOLOGIES

While there are a number of goals and potential benefits associated with federal loan guarantees for innovative clean energy technologies and projects, there are also multiple concerns about loan guarantees as an incentive mechanism for clean energy. The Congressional Budget Office (CBO) released a background paper in 1978 regarding concerns about loan guarantees for new energy technologies.

A brief overview of CBO's paper is provided in the text box at the end of this section.

Cash Flow Demand for Development-Stage Companies

A company that uses a loan guaranteed by the federal government to finance capital projects, or other business operations, has a legally binding requirement to pay back principal and interest to the loan issuer based on a defined repayment schedule.

Additionally, loan agreements typically have certain conditions and covenants that may require a company to maintain minimum cash holding levels for certain cash accounts.[33]

Therefore, a loan essentially results in a source of demand for a company's operating cash flow. For most development stage companies, managing cash flow is the essential financial management function that enables a company to operate and ultimately survive.

However, when development-stage companies with pre-commercial technologies use loans to finance new technology deployment (e.g., manufacturing facilities), the loan repayment requirements could potentially increase cash flow demands on a company and thus create liquidity challenges (see *Figure 7*).

The significant cash flow demands during this stage of a company's development could result in a high risk of loan default. Many companies in this development stage do not have an established commercial presence in their respective markets and are spending substantial amounts of cash to develop a sales force, establish marketing and distribution channels, complete technology performance validation, and establish other core elements of a sustainable business operation.

Additionally, many companies in this development stage will sell products at a loss as they work to achieve production economies of scale, which may or may not be realized. Using a loan as a means to finance a corporate asset, such as a manufacturing facility, during this development phase could potentially increase total cash flow demand and the likelihood of defaulting on the loan. In essence, loan guarantees may encourage the use of debt funding during risky development and deployment stages that might be more appropriate for equity investments.

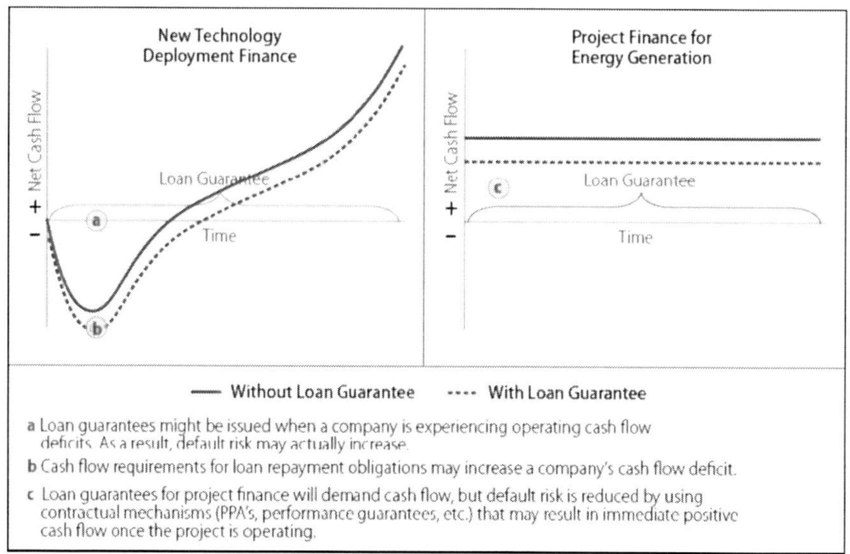

Source: CRS

Notes: Net Cash Flow refers to operating cash flow minus debt service requirements. Funds from loans for corporate assets are typically used to build and construct a particular asset. Companies might not be able to use funds for long term loans to support short term operating cash flow deficits. However, the company receiving the loan must be able to generate positive operating cash flow to fulfill its debt service obligations.

Figure 7. Illustrative Cash Flow Profiles for Clean Energy Technologies. (New technology deployment and project finance).

Solyndra, which received a loan guarantee for a manufacturing facility, might be considered an example of a new technology deployment project with high cash flow demands. Figure 8 shows Solyndra's actual operating losses from 2005 to 2009. Solyndra finalized its loan guarantee agreement in September 2009. Solar market conditions, which changed dramatically between 2009 and 2011, contributed to the company's negative operating cash flow during this period.34 Several reports indicate that when Solyndra initially defaulted on its loan obligation in 2010 the primary cause was due to cash flow issues that prevented the company from making a $5 million payment per the terms of the loan agreement.35 This example is not meant to show any cause and effect relationship between loan guarantees and bankruptcies or defaults. Rather, it illustrates the potential difficulty development stage companies might encounter when having to service debt obligations during periods of market uncertainty with high degrees of cash flow demand.

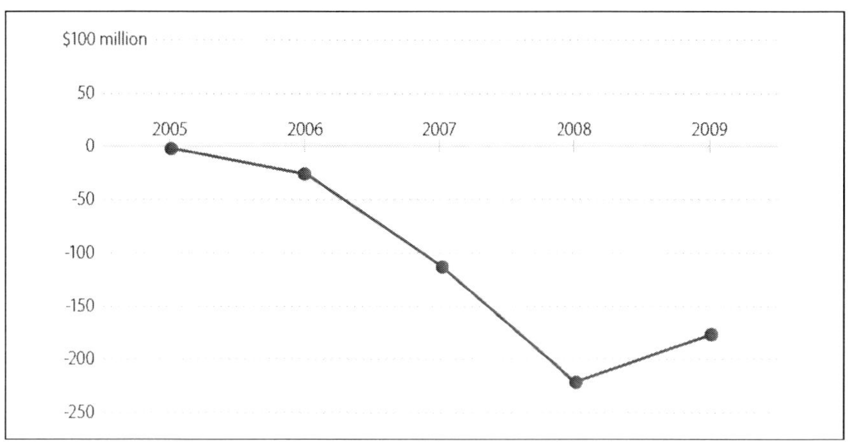

Source: Solyndra SEC filing, available at http://www.sec.gov/Archives/edgar/data/ 1443115/000119312510058567/ ds1a.htm#toc15203_8.

Figure 8. Actual Solyndra Operating Losses. (2005–2009).

On the other hand, using loan guarantees as a way to finance renewable electricity generation projects may be less risky since these types of projects are generally supported by long-term power purchase agreements (PPAs) and other contractual agreements that may provide a stable source of revenue and positive cash flow (see Figure 7). Since the risk profile of such projects might be low, some critics of clean energy loan guarantees may question why a federal loan guarantee is needed for these projects. However, default risk for these types of projects does exist and can result from technology performance and operational cost risks.

Indeed, different companies have different cash flow requirements, and cash management is best assessed on a project-by-project basis. Also, federally guaranteed loans may demand less cash when compared to commercial loans since interest rates on guaranteed loans are typically lower than those available in the commercial debt market. Nevertheless, a loan guarantee can still result in an additional cash flow burden for a company that is operating in the early stages of commercial deployment.

Government Risk/Reward Imbalance

Unlike corporate entities such as banks, private equity firms, and venture capital firms, the federal government is generally not designed to seek profits and financial returns. However, since taxpayer dollars are the source of federal financial incentive programs, when considering certain financial incentive policies it is worth considering how such policies will benefit the federal government, the country, and U.S. citizens. Loan guarantees are federal government commitments to fulfill the repayment obligations of certain loans in the event the borrower defaults. In essence, unless a loan guaranteed by the federal government defaults, the "cost" of the loan guarantee is essentially zero. However if a guaranteed loan defaults, then the federal government may be required to pay back principal and interest to the loan issuer, at which time the "cost" to the government could be as high as the total amount of principal borrowed for the loan.

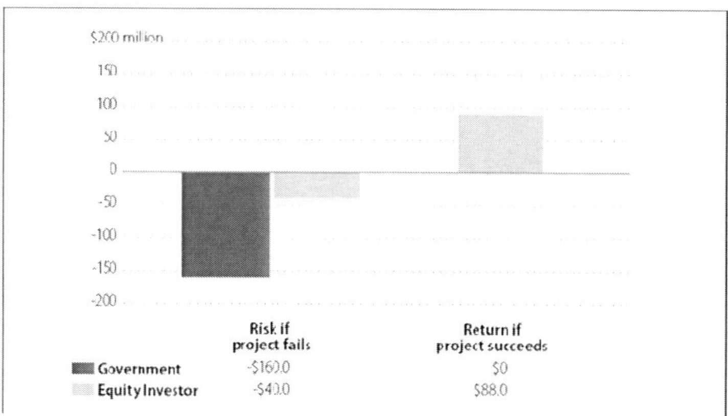

Source: CRS project finance analysis of a hypothetical solar module manufacturing facility. For a description of the model, project parameters, and financial assumptions used for this analysis, please contact the author directly.

Notes: It is important to note that the simplified example loan guarantee analysis in this figure is based on certain model input parameters and project finance assumptions. Using a different methodology, model inputs, and assumptions could produce very different results. The analysis is illustrative in nature and is not intended to predict real-world outcomes, which will differ based on actual project and market characteristics. Also, this analysis assumes that the example project defaults on its loan immediately following all loan disbursements. Losses to the government could be less if the project operated for a certain period of time, during which principal and interest payments were made. Numbers in this chart are on a Net Present Value basis.

Figure 9. Risk/Reward Profile for a Loan Guarantee Project Example: hypothetical solar module manufacturing project $200 million total project cost; 80% federal loan guarantee.

In financial terms, the federal government is risking an amount equal to the amount of principal guaranteed, yet the potential direct financial return to the government is essentially zero (See *Figure 9*). Financial return for the government is zero because the loan may be issued either by a commercial debt provider, who receives loan interest payments, or by the Federal Financing Bank (FFB).[36] FFB loans have low interest rates that are generally equal to Treasury debt.[37] Therefore, FFB is typically not making any money on an interest rate spread. Rather, FFB may use the interest received from federally guaranteed loans to pay down the Treasury debt used to source the loan funds.

The loan guarantee example illustrated in *Figure 9* does not take into account potential U.S. government benefits associated with job creation, a potentially larger tax base, and increased exports if the project succeeds. These benefits could be substantial, yet they are very difficult to accurately quantify and include in this type of analysis. As such, quantifying these potential benefits is beyond the scope of this report. Nevertheless, at the individual project level, some might perceive the government's risk/reward profile to be somewhat out of balance. Charging "credit subsidy costs" to projects that receive loan guarantees is one way the federal government attempts to mitigate the risk of losses associated with loan guarantees. However, under Section 1705, all credit subsidy costs for loan guarantees were paid for by appropriated funds. As a result, risk of loss to the Section 1705 Loan Guarantee Program is effectively reduced, yet the federal government is assuming all risks associated with loan defaults under the program.

Long-Term Commitments in a Dynamic Marketplace

Loans for renewable energy projects typically have a payback period of between 20 and 30 years, where the borrower is typically required to make periodic (monthly, quarterly) principal and interest payments based on terms and conditions of the loan agreement. Loan guarantees may cover the entire duration of a loan agreement. Especially for corporate finance activities that might support new technology manufacturing projects, the long-term nature of loans, and loan guarantees, is somewhat in contrast with the rapidly evolving renewable energy technology landscape. Innovation is occurring in the energy marketplace through venture capital investments in new energy technologies and federal government energy innovation programs. For example, the Department of Energy manages the SunShot Initiative, which "aims to dramatically decrease the total costs of solar energy systems by 75% before the end of the decade."[38] Successful future renewable energy technology innovations could, theoretically, make current technologies obsolete. As a result, technologies that may be commercially viable today could become outdated in less than a decade. The dynamic nature, and potential technology obsolescence, of renewable energy markets could introduce a certain amount of risk associated with using long-term loan guarantee commitments as an incentive mechanism for certain types of renewable energy projects. Furthermore, the amortized payback schedule of most debt instruments increases the risk to the government of principal losses associated with loan defaults that result from technology obsolescence.[39]

Pressure to Approve Loan Guarantees

A federally managed loan guarantee program for large clean energy projects essentially performs several banking-like functions. Financial analysis, market analysis, company due diligence, and other activities must be managed by such programs to facilitate sound financing decisions on the part of the federal government. However, government-managed loan guarantee efforts may be subject to certain pressures that might not be experienced by commercial banks. For example, Section 1705 was a temporary program, and loan guarantee authority under Section 1705 ended on September 30, 2011. Evaluation and proper due diligence of large, in some cases more than $1 billion, loan guarantee projects can take considerable amounts of time. Furthermore, there are certain project finance variables (executing power purchase agreements, supply agreements) that may not be within the immediate control of the Loan Programs Office. Therefore, having a pre-defined deadline for making loan guarantee commitments, along with a desire to expedite funding for technology deployment projects, may have adverse results. Projects that received loan guarantees may not be the best projects to have supported; rather these projects may have been in a better position to meet the deadlines associated with Section 1705 loan guarantee authority.

POLICY OPTIONS

Should Congress decide to debate the use of loan guarantees, or other government financial tools, as a clean energy deployment support mechanism, several policy options might be explored as a means to achieve clean energy policy objectives. As discussed earlier, a primary goal for loan guarantee programs is to provide a source of capital to projects that may not be able to secure low cost financing in the commercial market. Should this continue to be the fundamental objective of this type of incentive mechanism, the following discussion explores some policy options that Congress may also choose to consider.

Energy Project Loan Guarantee Concerns: Congressional Budget Office 1978

In August of 1978, the Congressional Budget Office (CBO) published a background paper titled: "Loan Guarantees: Current Concerns and Alternatives for Control." At that time, Congress had passed several laws that authorized the use of loan guarantees for energy projects. In its paper, CBO expressed several concerns about the use of loan guarantees for supporting such projects as well as concerns about the budgetary treatment of federal credit programs. As discussed previously in this report, the Federal Credit Reform Act (FCRA) was later enacted to improve budgetary treatment of direct loans and loan guarantees. However, other concerns outlined by CBO in its working paper may still be relevant today. Following is a brief discussion of some of CBO's concerns.

- Risk Evaluation by Lenders: When commercial lenders originate loans that are guaranteed by the government, these lenders may be more concerned with the adequacy of the loan guarantee agreement than by the actual risk of the project. As a result, projects may not receive an adequate amount of due diligence by the lender, therefore increasing the federal government's risk exposure.

- Partial Guarantees May Only Provide A Partial Solution: One way to improve lender risk evaluation is to require that lenders provide a certain portion of the loan principal in the form of a non-guaranteed loan. This would, in theory, increase the amount of scrutiny of loans by lenders. However, a small non-guaranteed loan requirement could potentially be absorbed by the lending organization by writing off losses against tax liabilities. Furthermore, the ability of lenders to securitize and sell the government-guaranteed portion of the loan could result in fees and returns that offset the risk of the non-guaranteed loan commitment.

> Furthermore, the goal of loan guarantee programs is to reduce the lender's risk. CBO highlights the fact that "while such guarantees reduce the risk of loss to lender and borrower, they cannot reduce the project's risk of economic failure." As a result, loan guarantees shift default risk from the lender and borrower to the federal government. The CBO paper also notes that loan guarantees are typically attractive to policy makers due to their perceived low cost. However, not truly understanding the full costs and effects of the federal government assuming long-term contingent liabilities could result in undesirable outcomes. The subsidy cost requirement, per FCRA, is a way to address full accounting for the true costs of loan guarantee programs. However, DOE's Section 1705 loan guarantee program is the largest amount of loan guarantees ever provided to support the deployment of innovative clean energy technologies. Only time will tell if subsidy cost estimates were adequate to compensate for actual losses associated with project defaults under the program.

Grants or Tax Expenditures Instead of Loan Guarantees

Grants for innovative clean energy technologies are a policy tool that could be used to incentivize commercialization and deployment of such technologies. Instead of appropriating funds to pay for loan guarantee subsidy costs, Congress could appropriate funds for a grant program that would provide financial assistance to projects that commercialize new energy technologies. The grant program could be structured in such a way that requires projects receiving federal grants to have secured all other necessary funding before receiving grant funds. Congress could also utilize tax expenditures as a financial mechanism for incentivizing the deployment of innovative clean energy technologies. Production tax credits and investment tax credits are two mechanisms currently used to incentivize renewable energy projects.[40] Companies receiving a federal grant or tax incentive would not be required to repay the grant amount or tax expenditure and, as a result, may not experience additional cash flow demands associated with loan repayments. In theory, using funds in this manner could be just as effective as using appropriated funds for subsidy costs. However, in practice different incentive mechanisms may be more useful depending on market characteristics and the financial credit environment. Using grants and tax expenditures as incentive mechanisms would limit the federal government's exposure to project failures. However, a drawback to this approach may be that using grants or tax expenditures, compared with loan guarantees, may not be perceived as providing an opportunity to leverage government funds.[41]

Equity Positions

One option Congress could explore is setting up a structure in which the federal government can assume equity positions in innovative clean energy technologies and projects. Since initial commercial deployment of new technologies is high risk in nature, equity investments, arguably, might be more appropriate than loans or loan guarantees for this stage of the technology commercialization life cycle. Equity positions might serve to alleviate the cash flow demands associated with loans and may also provide the federal government with an opportunity to participate in the return upside if a project is successful. Thus, equity positions in clean energy technologies may serve to balance the federal government's risk/return profile. Making these types of high risk investments may require the federal government to operate much like a venture capital firm, where a portfolio of equity positions are taken in high risk/high return investments. The overall goal would be that successful projects should more than compensate for project failures. Congress could create a clean energy venture capital-like entity that would have the funding, charter, and authority needed to invest in commercial deployment of innovative clean energy technologies. However, this approach raises concerns about the federal government assuming a venture capital-like function and how such an organization may improve or hinder the existing venture capital and private equity community. Furthermore, equity positions in companies also raise concerns about the federal government control of industry. However, federal government equity positions are not unprecedented. Financial support in return for such positions has been provided recently to auto companies, banks, and others. In those instances, this type of financial assistance was done under what might be considered emergency circumstances and not without controversy.[42]

Flexible Financial Management Tools

Should Congress decide to continue using loan guarantees as a support mechanism for clean energy deployment, providing authority for loan programs to use certain flexible financial management tools may be an option to consider. Financial tools that might be used by federal loan programs may include the following:

- *Warrants:* A stock warrant provides the holder of that warrant the opportunity to purchase a company's stock at a certain price sometime in the future. As part of a loan guarantee agreement, the federal government could possibly receive warrants from companies that receive loan guarantees. These warrants would provide the federal government with an opportunity to participate in the financial return of successful projects and balance the risk/return profile of individual projects. The use of warrants could be a way for the federal government to recover appropriated credit subsidy costs used for loan guarantee projects.[43]

- *Portfolio management:* Portfolio management is intended to ensure that gains from certain projects would offset, and possibly exceed, losses from other projects. Currently, innovative clean energy technology loan guarantees are managed on a project-by-project basis and there is no opportunity to reduce the risk of losses through portfolio management. A portfolio management approach, along with financial tools such as warrants, may serve to reduce the overall financial risk of loan guarantee programs.

- *Convertible preferred equity:* To reduce the initial cash flow demands associated with loans and loan guarantees, Congress might consider the use of a convertible preferred equity instrument as a way to fund innovative clean energy projects. The concept would be, for example, for the federal government to provide the necessary funding needed for a new project and, in return, receive a controlling preferred equity position in the project or company.[44] Once the project, or company, has achieved positive cash flow that would allow for adequate debt service, the preferred equity is converted into debt, which is then repaid based on a determined repayment schedule. This approach would give the federal government a high degree of management control of the project during its startup phase, a clear incentive for the project/company to realize positive cash flow as soon as possible, and a reasonable loan repayment schedule to recover the investment. Furthermore, this approach may reduce cash flow demand during the initial start-up phase of projects. Although, as discussed in the "Equity Positions" section above, this approach raises concerns about the level of federal government control.

Clean Energy Financial Support Authority

Should Congress decide to continue supporting development and deployment of clean energy technologies, creating an organization to manage various forms of federal financial support for such endeavors may be a policy option to consider. The organization could be given authority to utilize various financial tools to manage a portfolio of clean energy deployment investments. This new organization could be located within an existing federal agency or it could be an independent body. If Congress were to decide to locate this new organization within an existing federal agency, it may want to evaluate the most appropriate federal agency to be chosen. The Department of Energy is where the current clean energy deployment loan guarantee program resides and DOE may be the appropriate agency for such a program. However, Congress may want to consider the U.S. Treasury as another option for locating a new clean energy financing authority as Treasury may offer existing finance, banking, and investment expertise that could potentially manage an organization with a variety of financial investment tools.

LEGISLATIVE ACTION

In the 112th Congress, the Clean Energy Financing Act of 2011 (S. 1510) proposes to create a Clean Energy Deployment Administration (CEDA) within the Department of Energy. As proposed in S. 1510, CEDA would be able to use financial tools such as direct loans, loan guarantees, and insurance products to support clean energy technology manufacturing and deployment. The bill allows for a portfolio management approach as a way to manage financial risk. S. 1510 also allows the use of "alternative fee arrangements" such as profit participation, stock warrants, and others as a way to potentially reduce the amount of upfront cash fees.

End Notes

[1] For more information about the Solyndra bankruptcy, see CRS Report R42058, *Market Dynamics That May Have Contributed to Solyndra's Bankruptcy*, by Phillip Brown.

[2] Congressional Budget Office, "Loan Guarantees: Current Concerns and Alternatives for Control," August 1978.

[3] Office of Management and Budget, "Analytical Perspectives, Budget of the United States Government, Fiscal Year 2012," Table 23-12 Guaranteed Loan Transactions of the Federal

Government, available at http://www.whitehouse.gov/sites/default/files/omb/budget/fy2012/assets/23_12.pdf.

[4] OMB Circular A-11, November 2011.

[5] Congressional Budget Office, "Loan Guarantees: Current Concerns and Alternatives for Control," August 1978.

[6] For more information about FCRA, see CRS Report RL30346, Federal Credit Reform: Implementation of the Changed Budgetary Treatment of Direct Loans and Loan Guarantees, by James M. Bickley.

[7] Specific OMB credit subsidy guidance is as follows: "The subsidy cost is the estimated present value of the cash flows from the Government (excluding administrative expenses) less the estimated present value of the cash flows to the Government resulting from a direct loan or loan guarantee, discounted to the time when the loan is disbursed. The cash flows are the contractual cash flows adjusted for expected deviations from the contract terms (delinquencies, defaults, prepayments, and other factors)." For more information about OMB guidance for calculating credit subsidy costs, see OMB Circular No. A-11, Part 5— Federal Credit, November 2011, available at http://www.whitehouse.gov/ sites/default/files/omb/assets/a11_current_year/s185.pdf.

[8] Congressional Budget Office, "Loan Guarantees: Current Concerns and Alternatives for Control," August 1978.

[9] Ibid.

[10] Mark Holt, "Energy policy: Is the U.S. ready for the 1990s?" Environmental and Energy Study Conference, April 18, 1988.

[11] U.S. Government Accounting Office, "Financial Status of the Great Plains Coal Gasification Project," February 21, 1985, available at http://archive.gao.gov/d10t2/126322.pdf.

[12] U.S. Government Accounting Office, "Status of the Great Plains Coal Gasification Project," November 2005, available at http://www.gao.gov/assets/90/86915.pdf.

[13] Terms of the purchase agreement included revenue sharing that required Dakota Gasification to provide cash payments to DOE when natural gas sales prices exceeded a certain level. According to Dakota Gasification, $391 million was paid to DOE through 2009 when the revenue sharing requirement expired. For more information see http://www.dakotagas.com/About_Us/Finance/index.html.

[14] http://www.dakotagas.com/About_Us/History/1989-Present/index.html.

[15] Mark Holt, "Energy policy: Is the U.S. ready for the 1990s?" Environmental and Energy Study Conference, April 18, 1988.

[16] USDA manages a loan guarantee program, the Biorefinery Assistance Program, to assist emerging renewable transportation fuel production technologies. USDA's loan guarantee program is not the focus of this report. However, more information about the Biorefinery Assistance Program is available at http://www.rurdev.usda.gov/ BCP_Biorefinery.html.

[17] The DOE Loan Guarantee Program was initially operated by DOE's Office of the Chief Financial Officer. The Loan Guarantee Program was later merged with the Advanced Technology Vehicle Manufacturing loan program to form DOE's Loan Programs Office.

[18] Loan guarantee projects that receive funds from the Federal Financing Bank (FFB) are classified as "Direct Loans" in the federal budget. Most funds for 1705 loan guarantee recipients came from the FFB and are in the "Direct Loan" budget category. It was therefore necessary to include direct loans and loan guarantees for this analysis in order to accurately estimate the relative magnitude of DOE's loan guarantee program activities.

[19] Section 1701 of EPACT 2005 provides the following definition of commercial technology: "The term 'commercial technology' means a technology in general use in the commercial marketplace."

[20] FCRA Section 502(5)(C) provides the following definition for loan guarantee cost: "The cost of a loan guarantee shall be the net present value, at the time when the guaranteed loan is disbursed, of the following estimated cash flows: (i) payments by the Government to cover defaults and delinquencies, interest subsidies, or other payments; and (ii) payments to the Government including origination and other fees, penalties and recoveries; including the effects of changes in loan terms resulting from the exercise by the guaranteed lender of an option included in the loan guarantee contract, or by the borrower of an option included in the guaranteed loan contract."

[21] Appendix to the Budget of the United States Government—Fiscal Year 2012, Department of Energy detailed budget estimate, Office of Management and Budget, available at http://www.whitehouse.gov/sites/default/files/omb/budget/ fy2012/assets/doe.pdf.

[22] More information about DOE's Loan Programs Office is available at http://lpo.energy.gov/.

[23] ATVM is a direct loan program and is not discussed in detail within this report. For more information about DOE's ATVM program see CRS Report R42064, The Advanced Technology Vehicles Manufacturing (ATVM) Loan Program: Status and Issues, by Brent D. Yacobucci and Bill Canis.

[24] Section 1703 conditional loan guarantee commitments are dominated by two nuclear electricity generation projects valued at $2 billion and $8.33 billion, respectively. For more information see https://lpo.energy.gov/?page_id=45.

[25] For a detailed comparison of manufacturing and electricity generation projects see CRS Report R42059, Solar Projects: DOE Section 1705 Loan Guarantees, by Phillip Brown.

[26] DOE LPO Section 1705 supported a number of solar electricity generation projects. However, technologies these projects include commercially available solar photovoltaic modules as well as various concentrating solar thermal technologies. Solar thermal technologies might generally be considered less commercially available than solar PV technologies and, as a result, the technology performance risk of projects that use solar thermal technologies might be relatively high.

[27] For more information about commercialization challenges and potential strategies to address certain challenges, see Geoffrey A. Moore, Crossing the Chasm: Marketing and Selling High-Tech Products to Mainstream Customers (HarperCollins, 2002).

[28] International Energy Agency (2011), World Energy Outlook 2011, OECD Publishing.

[29] Ibid.

[30] Department of Energy Loan Programs Office, https://lpo.energy.gov/?page_id=45.

[31] https://lpo.energy.gov/?page_id=45.

[32] For information about the U.S. wind manufacturing supply chain, see CRS Report R42023, U.S. Wind Turbine Manufacturing: Federal Support for an Emerging Industry, by Michaela D. Platzer.

[33] For example, loan guarantee agreements may require recipients to maintain minimum balances in reserve accounts for debt service, operations and maintenance, among others.

[34] For more information see CRS Report R42058, Market Dynamics That May Have Contributed to Solyndra's Bankruptcy, by Phillip Brown.

[35] Deborah Solomon, "Solyndra Said to Have Violated Terms of Its U.S. Loan," Wall Street Journal, September 28, 2011.

[36] For more information about the Federal Financing Bank, see the FFB website at http://www.treasury.gov/ffb/ index.shtml.

[37] FFB may charge a small premium of 1/8th of 1% to cover charges associated with servicing loans.

[38] For more information about DOE's SunShot Initiative, see http://www1.eere.energy.gov/solar/sunshot/about.html.

[39] Loans are typically amortized over a certain number of years. Generally speaking, initial debt service payments usually include more interest than principal in the early years and more principal than interest in later years. For the loan guarantee analysis described in Figure 9, roughly 38% of the debt principal is repaid after 10 years (loan tenor for the example analysis is 20 years). Therefore, approximately 62% of the principal will be repaid in the second half of the project.

[40] For more information see CRS Report R41227, Energy Tax Policy: Historical Perspectives on and Current Status of Energy Tax Expenditures, by Molly F. Sherlock.

[41] An example of the potential leverage opportunity associated with loan guarantees is DOE's Section 1705 program. Approximately $2.5 billion, after rescissions and transfers, of credit subsidy costs were appropriated and the program supported a loan value of approximately $16.15 billion.

[42] The federal government provided financial assistance in return for equity positions as part of the Troubled Asset Relief Program (TARP), for more information see CRS Report R41427, Troubled Asset Relief Program (TARP): Implementation and Status, by Baird Webel.

[43] Typically warrant holders are not entitled to seats on a company's board of directors, and are not able to vote on corporate affairs issues.

[44] Preferred equity may also have certain rights such as dividend preferences as well as common stock conversion multiples. Preferred equity rights vary on a company-by-company basis.

Chapter 2

SOLAR PROJECTS:
DOE SECTION 1705 LOAN GUARANTEES[*]

Phillip Brown

INTRODUCTION

Since Solyndra, a solar system manufacturing company that received a $535 million loan guarantee from the Department of Energy (DOE), filed for bankruptcy in September of 2011 there has been much congressional interest in better understanding the characteristics of renewable energy projects, specifically solar projects, that have received DOE loan guarantees. The objective of this report is to provide Congress with insight regarding solar projects supported by DOE's loan guarantee program, the risk characteristics of these projects, and how other DOE loan guarantee projects are either similar to or different from the Solyndra solar manufacturing project.

KEY POINTS

- DOE's Loan Programs Office (LPO) administers three separate loan programs: (1) Section 17031 loan guarantees, (2) Section 17052 loan

---

[*] This is an edited, reformatted and augmented version of a Congressional Research Service publication, CRS Report for Congress R42059, from www.crs.gov, prepared for Members and Committees of Congress, dated October 25, 2011.

guarantees, and (3) Advanced Technology Vehicle Manufacturing (ATVM) loans.

- To date, all loan guarantees for solar projects have been provided under LPO's Section 1705 program.

- LPO's Section 1705 program has closed transactions that guarantee approximately $16.15 billion of loans for renewable energy projects. Roughly 82% ($13.27 billion) of Section 1705 loan guarantees have been for solar projects.

- Solar projects supported by Section 1705 loan guarantees generally fall into one of two categories: (1) solar manufacturing, or (2) solar generation. Each category has different financial, operational, and technology risk characteristics.

- Four solar manufacturing projects, including Solyndra, have received loan guarantees totaling $1.28 billion, which is approximately 8% of the total dollar value of loans guaranteed under the Section 1705 program.

- Twelve solar generation projects have received loan guarantees totaling $11.99 billion, which is approximately 74% of the total dollar value of loans guaranteed under the Section 1705 program.

- Solar manufacturing projects might generally be considered more risky than solar generation projects because solar generation projects have contractual mechanisms (power purchase agreements, performance guarantees, service agreements, etc.) that allow these projects to manage many project financial risks.

- One solar manufacturing project might be considered somewhat similar to Solyndra, only because the project aims to manufacture solar panels that use the same materials as those used by Solyndra (copper indium gallium selenide— CIGS). However, the company's manufacturing approach, products, and markets are distinctly different from those of Solyndra.

- All LPO solar manufacturing projects will have to address and manage the same market risks that may have contributed to Solyndra's bankruptcy. These risks include (1) declining solar module prices; (2) competition from new and established solar manufacturing competitors; and (3) subsidy/incentive reductions in international (mostly European) markets.

BACKGROUND—DOE LOANS PROGRAM

The Department of Energy Loan Programs Office (LPO) administers three loan programs:

- *Section 1703:* loan guarantees for innovative clean energy technologies with high degrees of technology risk.
- *Section 1705:* loan guarantees for certain renewable energy systems, electric power transmission, and innovative biofuel projects that may have varying degrees (high or low) of technology risk.
- *Advanced Technology Vehicle Manufacturing (ATVM):* loans to support advanced technology vehicles and associated components.

According to LPO's website, all renewable energy manufacturing and electricity generation projects supported by loan guarantees have used the Section 1705 program.[3] DOE's loan guarantee authority originated from Title XVII of the Energy Policy Act of 2005 (P.L. 109-58). The American Recovery and Reinvestment Act of 2009 (P.L. 111-5) amended the Energy Policy Act of 2005 by adding Section 1705. Section 1705 was created as a temporary program, and 1705 loan guarantee authority ended on September 30, 2011. DOE received appropriated funds to pay for credit subsidy costs[4] associated with Section 1705 loan guarantees, which, after rescissions and transfers, was $2.435 billion.[5] From an industry perspective, Section 1705 loan guarantees were very attractive as they provided an opportunity to obtain low-cost capital with the required credit subsidy costs paid for by appropriated government funds.[6]

LPO has guaranteed approximately $16.15 billion of loans for renewable energy projects. Roughly 82% ($13.27 billion) of the value of Section 1705 loan guarantees support solar projects (see *Figure 1*). The remaining 18% of Section 1705 loan guarantees support a variety of projects for biofuel production, energy storage, wind generation, transmission, and geothermal electricity. The focus of this report is on solar projects supported by Section 1705.

Solar projects supported by Section 1705 loan guarantees generally fall into one of two categories: (1) solar manufacturing, or (2) solar generation. Each project category typically has different market, customer, financial, and technology risk characteristics. Solyndra is categorized as a solar manufacturing project. The following sections discuss the different characteristics and risk profiles for the two project categories.

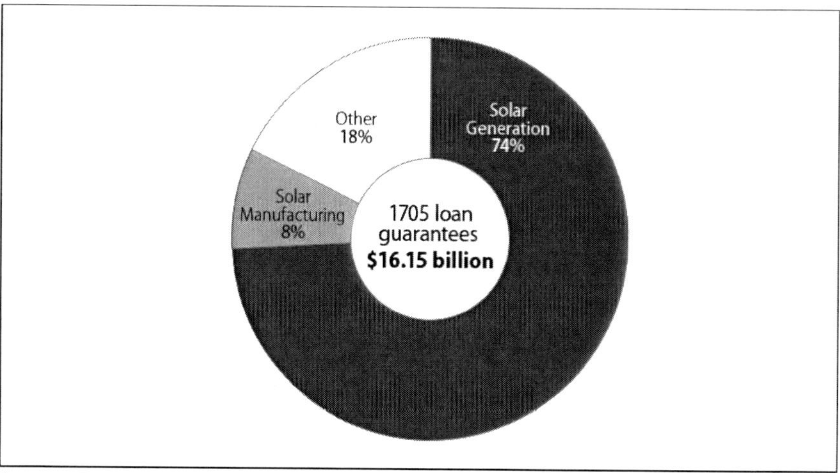

Source: Department of Energy Loan Programs Office website, https://lpo.energy. gov/?page_id=45.

Notes: "Other" projects include biofuel production, energy storage, wind generation, transmission, and geothermal electricity.

Figure 1. DOE Section 1705 Loan Guarantees.

SOLAR MANUFACTURING PROJECTS

The Department of Energy has committed to guarantee four loans totaling $1.28 billion for solar manufacturing projects, approximately 8% of the value of all loans guaranteed under Section 1705 (see *Table 1*). Solar manufacturing projects supported by 1705 are generally for scaling up manufacturing capacity for new solar technologies that may offer cost, performance, or other discriminators in the solar marketplace. Solyndra falls into this category.

Solar manufacturing projects typically have to manage several market risks in order to succeed. Each project must successfully scale-up its manufacturing capacity; produce technologies and products that are differentiated based on cost, performance, and/or other parameters; compete with companies that offer commercially proven products; and navigate a dynamic marketplace that has experienced dramatic cost reductions and new market entrants over the last several years.

Table 1. Solar Manufacturing Projects Supported By DOE's 1705 Loan Guarantee Program

Project	Loan Guarantee Amount	Technology
1366 Technologies	$150 million	Silicon solar wafer manufacturing process that may reduce silicon waste by as much as 50% compared with current processes.
Abound Solar	$400 million	Proprietary manufacturing process for thin-film cadmium telluride (CdTe) photovoltaic modules.
SoloPower	$197 million	Copper indium gallium selenide (CIGS) photovoltaic cell and module manufacturing using a proprietary electrochemical fabrication process.
Solyndra Inc.	$535 million	Cylindrical CIGS photovoltaic cell and module manufacturing for commercial rooftop applications.
Total	$1,282 million	

Source: DOE loan guarantee program website, https://lpo.energy. Company web sites.

One solar manufacturing project (SoloPower) might possibly be considered somewhat similar to Solyndra, only because the project aims to manufacture solar panels that use the same materials as those used by Solyndra (copper indium gallium selenide—CIGS). However, SoloPower's manufacturing approach (electrochemical), products (thin-film PV on a flexible substrate), and markets (residential and commercial rooftops) are distinctly different from those of Solyndra.

Nevertheless, each LPO-supported solar manufacturing project will have to address and manage the same market risks that may have contributed to Solyndra's bankruptcy.

These risks include (1) declining solar module prices; (2) competition from new and established solar panel manufacturers; and (3) subsidy/incentive reductions in international (mostly European) markets. The success or failure of each respective project will likely be determined by the ability of each solar manufacturing project to differentiate its product in the solar marketplace, deliver expected cost and performance objectives, and convince buyers to accept some degree of new technology risk.

SOLAR GENERATION PROJECTS

LPO has guaranteed twelve loans totaling $11.99 billion for solar generation projects, approximately 74% of loans guaranteed under the Section 1705 program (see *Table 2*).

These projects are generally large solar electricity generation projects that use commercially proven technologies or technologies that have some operational history from pre-commercial demonstrations.

Table 2. Solar Generation Projects Supported By DOE's 1705 Loan Guarantee Program

Type	Project (Company)	Loan Guarantee Amount	Technology
CSP	Mojave Solar (Abengoa)	$1,200 million	Parabolic trough concentrating solar power.
CSP	Solana (Abengoa)	$1,466 million	Parabolic trough concentrating solar power.
CSP	Genesis Solar (NextEra)	$852 million*	Parabolic trough concentrating solar power.
CSP	Crescent Dunes (Solar Reserve)	$737 million	Power tower concentrating solar power with thermal storage system.
CSP	Ivanpah (Brightsource)	$1,600 million	Power tower concentrating solar power.
CPV	Cogentrix	$90.6 million	Concentrating photovoltaic electricity generation technology.
PV	Antelope Valley (Exelon)	$646 million	FirstSolar thin film solar panels along with a new type of inverter technology.
PV	Mesquite Solar 1 (Sempra)	$337 million	Crystalline silicon photovoltaic solar modules from
			Suntech.
PV	Desert Sunlight (NextEra)	$1,460 million*	FirstSolar thin film solar panels.
PV	California Valley Solar Ranch	$1,237 million	Crystalline silicon photovoltaic solar modules from
	(NRG Energy)		SunPower.
PV	Agua Caliente (NRG Solar)	$967 million	FirstSolar thin film solar panels.
PV	Project Amp (Prologis)	$1,400 million*	Rooftop photovoltaic panels on 750 warehouse buildings owned by ProLogis, Inc.
	Total	$11,993 million	

Source: DOE loan guarantee program website, https://lpo.energy

Notes: CSP = concentrating solar power; CPV = concentrating photovoltaic; PV = photovoltaic. * DOE provided a partial guarantee.

Unlike solar manufacturing projects, solar generation projects generally do not have to deal with market risks. Rather, these projects typically have to manage operation and execution risks. One specific risk associated with these solar generation projects is their ability to perform and operate at large scale. Many of the power generation projects that have received loan guarantees use technologies that have been demonstrated either commercially or at a pilot scale.

However the size of each project is quite large and in some cases will be the largest project either nationally or globally to use its respective technology.

Most solar generation projects reduce market risk through the use of power purchase agreement (PPA) contracts with entities (often electric utility companies) that agree to purchase a project's electricity at defined prices for terms usually between 20 and 25 years.

As a result, the focus shifts to risks associated with performance of the selected technology over the contract period (generating enough electricity to meet revenue objectives) and the operations and maintenance (O&M) costs of the project during its lifetime (keeping O&M costs low to enable cash flows that can adequately service debt and other financial obligations). However, performance guarantees and O&M service agreements are typically used to manage these project-level risks.[7]

Furthermore, it is important to recognize that three different technology types are represented by the twelve solar generation projects supported by 1705 loan guarantees.

As indicated in *Table 2*, the three technology types include (1) photovoltaic (PV); (2) concentrating solar power (CSP); and (3) concentrating photovoltaic (CPV). PV technology is generally considered to be the most commercially viable of the three technology types, as the overwhelming majority of global solar installations use PV.

CSP technologies might be considered less commercial when compared to PV, although parabolic trough CSP systems do have some commercial operating history. CPV technology might be considered the least commercially viable of the three technologies, and the Cogentrix project supported by 1705, at 30 megawatts of generating capacity, will be the largest utility-scale CPV project in the world.[8]

Generally speaking, projects that use technologies that may not be fully commercialized (CSP and CPV) will be higher risk than projects that use commercially proven technology (PV).

RISK PROFILE: SOLAR MANUFACTURING VS. SOLAR GENERATION

For the purpose of this analysis, risk might be defined as dynamic external (market, competition, cost, etc.) and internal (technology, performance, operation) factors that can impact a project's ability to meet its financial obligations. It is important to recognize how risk characteristics of solar manufacturing and solar generation projects are different. Solar manufacturing projects typically have to manage several types of risk: market, technology, competitor, and others. Solar generation projects, on the other hand, generally need to only manage operation and execution risks. *Table 3* provides a side-by-side comparison of risks for solar manufacturing and solar generation projects. Solar manufacturing projects might generally be considered more risky than solar generation projects because solar generation projects can use contractual mechanisms to manage many project financial risks.

Table 3. Risk Comparison: Solar Manufacturing and Solar Generation Projects

	Solar Manufacturing	Solar Generation
Market risk	Projects must constantly manage changing market dynamics such as demand levels, manufacturing capacity levels, subsidies/incentives, etc.	Generally not an issue once the project is operating. Most generation projects reduce market risk through the use of power purchase agreements (PPA) for electricity generated.
Competition risk	Projects must monitor and respond to competitors with regard to technology improvements, cost reductions, and other factors.	Once a PPA is established with an electricity customer, competition risk for the project is reduced.
Cost risk	Solar manufacturing projects must keep pace with industry cost trends in order to maintain competitiveness in the marketplace.	To some degree, generation projects must manage operation and maintenance (O&M) costs in order to achieve financial projections. O&M cost risks might be managed through the use of O&M service agreements.
Technology/performance risk	Projects must ensure that the new and innovative technologies being manufactured are able to meet or exceed technology performance targets.	Many generation projects use commercially proven technologies that have minimum technology risk. Some 1705 projects are using technologies that are not fully

	Solar Manufacturing	Solar Generation
		commercialized, which may present some degree of technology performance risk. Technology and performance risks can be managed through performance guarantee agreements.
Customer acceptance risk	Solar manufacturing projects must actively work to convince potential customers to accept the new and innovative technology being manufactured. To some degree, this can be managed through sales agreements. However, in some cases sales agreements may not be contractually binding.	For projects that have established power purchase agreements, the customer acceptance risk is typically eliminated. Unlike solar manufacturing projects, solar generation project customers are purchasing electricity, not a particular technology.
Operation/execution risk	Projects must execute their construction and scale-up plans as well as their operational plans in order to achieve revenue and cost targets.	Generation projects have some degree of operation and execution risk as they need to operate in a manner that achieves revenue and cost targets. Further, many 1705 projects are the largest in either the country or the world for their respective technology types. The large-scale nature of these projects creates uncertainties that can result in operation and execution risk.

Source: CRS.

CONCLUSION

Each solar project supported by Section 1705 has a unique set of project, technology, and risk characteristics that require constant management. It is important to recognize, however, that Section 1705 solar projects fall into one of two categories: (1) solar manufacturing, or (2) solar generation. Risk characteristics for each category are distinctly different and solar manufacturing projects are generally considered higher risk than solar generation projects because the latter can use contractual mechanisms to reduce market, project, and financial risks. Whether or not Section 1705 solar projects will succeed is beyond the scope of this report. However, each Section 1705 solar manufacturing project will have to address the same market dynamics that may have contributed to Solyndra's bankruptcy. Ultimately, the success or failure of all Section 1705 solar projects will likely be determined by the ability of each project's management team to adapt to market dynamics and manage project risks.

End Notes

[1] Section 1703 of the Energy Policy Act of 2005 (P.L. 109-58).

[2] Section 1705 of the Energy Policy Act of 2005 (P.L. 109-58). Section 1705 was created by amending the Energy Policy Act of 2005 through the American Recovery and Reinvestment Act of 2009 (P.L. 111-5).

[3] Department of Energy Loans Program Office website, https://lpo.energy.gov/?page_id=45.

[4] DOE's LPO provides the following definition of credit subsidy cost: "Credit Subsidy Cost has the same meaning as 'cost of a loan guarantee' in section 502(5)(C) of the Federal Credit Reform Act of 1990 (2 U.S.C. 661a(5)(C)), which is the net present value, at the time the Loan Guarantee Agreement is executed, of the following estimated cash flows, discounted to the point of disbursement: (1) Payments by the Government to cover defaults and delinquencies, interest subsidies, or other payments; less (2) Payments to the Government including origination and other fees, penalties, and recoveries including the effects of changes in loan or debt terms resulting from the exercise by the Borrower, Lead Lender or other Holder of an option included in the Loan Guarantee Agreement." See https://lpo.energy.gov/?page_id= 64.

[5] Department of Energy 2012 budget appendix, available at http://www.whitehouse.gov /sites/default/files/omb/budget/ fy2012/assets/doe.pdf.

[6] Section 1703 loan guarantees required the loan guarantee recipient to pay the credit subsidy cost.

In: Federal Loan Guarantees for Energy Projects   ISBN: 978-1-62417-116-1
Editors: R. Sampson and C. Lange   © 2013 Nova Science Publishers, Inc.

Chapter 3

DOE LOAN GUARANTEES: FURTHER ACTIONS ARE NEEDED TO IMPROVE TRACKING AND REVIEW OF APPLICATIONS[*]

The United States Government Accountability Office

ABBREVIATIONS

CRB	Credit Review Board
DOE	Department of Energy
EERE	Energy Efficiency, Renewable Energy
FIPP	Financial Institution Partnership Program
LGP	Loan Guarantee Program
LPO	Loan Programs Office
OMB	Office of Management and Budget

---

[*] This is an edited, reformatted and augmented version of The United States Government Accountability Office publication, Report to Congressional Committees, GAO-12-157, dated March 2012.

WHY GAO DID THIS STUDY

The Department of Energy's (DOE) Loan Guarantee Program (LGP) was created by section 1703 of the Energy Policy Act of 2005 to guarantee loans for innovative energy projects. Currently, DOE is authorized to make up to $34 billion in section 1703 loan guarantees. In February 2009, the American Recovery and Reinvestment Act added section 1705, making certain commercial technologies that could start construction by September 30, 2011, eligible for loan guarantees. It provided $6 billion in appropriations that were later reduced by transfer and rescission to $2.5 billion. The funds could cover DOE's costs for an estimated $18 billion in additional loan guarantees. GAO has an ongoing mandate to review the program's implementation. Because of concerns raised in prior work, GAO assessed (1) the status of the applications to the LGP and (2) for loans that the LGP has committed to, or made, the extent to which the program has adhered to its process for reviewing applications. GAO analyzed relevant legislation, regulations, and guidance; prior audits; and LGP data, documents, and applications. GAO also interviewed DOE officials and private lenders with experience in energy project lending.

WHAT GAO RECOMMENDS

GAO recommends that the Secretary of Energy establish a timetable for, and fully implement, a consolidated system to provide information on LGP applications and reviews and regularly update program policies and procedures. DOE disagreed with the first of GAO's three recommendations; GAO continues to believe that a consolidated system would enhance program management.

WHAT GAO FOUND

The Department of Energy (DOE) has made $15 billion in loan guarantees and conditionally committed to an additional $15 billion, but the program does not have the consolidated data on application status needed to facilitate efficient management and program oversight. For the 460 applications to the Loan Guarantee Program (LGP), DOE has made loan guarantees for 7 percent and committed to an additional 2 percent. The time the LGP took to review loan applications decreased over the course of the program, according to GAO's analysis of LGP data. However, when GAO requested data from the LGP on the status of these applications, the LGP did not have consolidated data readily available and had to assemble these data over several months from various sources. Without consolidated data on applicants, LGP managers do not have readily accessible information that would facilitate more efficient program management, and LGP staff may not be able to identify weaknesses, if any, in the program's application review process and approval procedures. Furthermore, because it took months to assemble the data required for GAO's review, it is also clear that the data were not readily available to conduct timely oversight of the program. LGP officials have acknowledged the need for a consolidated system and said that the program has begun developing a comprehensive business management system that could also be used to track the status of LGP applications. However, the LGP has not committed to a timetable to fully implement this system.

The LGP adhered to most of its established process for reviewing applications, but its actual process differed from its established process at least once on 11 of the 13 applications GAO reviewed. Private lenders who finance energy projects that GAO interviewed found that the LGP's established review process was generally as stringent as or more stringent than their own. However, GAO found that the reviews that the LGP conducted sometimes differed from its established process in that, for example, actual reviews skipped applicable review steps. In other cases, GAO could not determine whether the LGP had performed some established review steps because of poor documentation. Omitting or poorly documenting reviews reduces the LGP's assurance that it has treated applicants consistently and equitably and, in some cases, may affect the LGP's ability to fully assess and mitigate project risks. Furthermore, the absence of adequate documentation may make it difficult for DOE to defend its decisions on loan guarantees as sound and fair if it is questioned about the justification for and equity of those decisions. One cause of the differences between established and actual processes was that, according to LGP staff, they were following procedures that had been revised but were not yet updated in the credit policies and procedures manual, which governs much of the LGP's established review process. In particular, the version of the manual in use at the time of GAO's review was dated March 5, 2009, even though the manual states it was meant to be updated at least annually, and more frequently as needed. The updated manual dated October 6, 2011, addresses many of the differences GAO identified. Officials also demonstrated that LGP had taken steps to address the documentation issues by beginning to implement its new document management system. However, by the close of GAO's review, LGP could not provide sufficient documentation to resolve the issues identified in the review.

March 12, 2012

The Honorable Dianne Feinstein
Chair

The Honorable Lamar Alexander
Ranking Member
Subcommittee on Energy and Water Development
Committee on Appropriations
United States Senate

The Honorable Rodney P. Frelinghuysen
Chairman

The Honorable Peter J. Visclosky
Ranking Member
Subcommittee on Energy and Water Development,
and Related Agencies
Committee on Appropriations
House of Representatives

The Department of Energy's (DOE) loan guarantee program (LGP) is currently authorized to issue loan guarantees worth up to $34 billion for certain types of energy projects that need affordable financing.[1] Federal loan guarantee programs such as the LGP can help companies obtain such financing because the federal government agrees to reimburse the lender for the guaranteed amount if a borrower defaults. As directed by section 1703 of the Energy Policy Act of 2005, the LGP originally focused on projects that use new or significantly improved energy technologies and avoid, reduce, or sequester emissions of air pollutants or man-made greenhouse gases. In February 2009, Congress expanded the scope of the LGP in the American Recovery and Reinvestment Act (Recovery Act), by adding section 1705 to the Energy Policy Act, which provided funding and extended the program to include projects that use commercial energy technology that employs renewable energy systems, electric power transmission systems, or leading-edge biofuels that meet certain criteria. The LGP has issued nine calls for applications—known as solicitations— each of which covers particular types of energy technology.

According to DOE officials, the LGP is important to both develop new energy technologies for commercial use and make some commercial projects possible, thereby creating jobs and new energy supplies. However, loan guarantee programs can also expose the government to substantial financial risks. For example, a borrower could default on a federally guaranteed loan, leaving taxpayers to pay for the loss. In the past, we also found problems with federal loan guarantee programs that occurred in part because agencies did not exercise sufficient due diligence. Due diligence is the review process by which a lender identifies and mitigates potential problems or risks with a project before the lender makes a loan or loan guarantee. Recently, the filing of bankruptcy petitions by two recipients of DOE loan guarantees have raised concerns that DOE may not be sufficiently identifying and mitigating the risk of a loan default.

GAO has an ongoing mandate under the 2007 Revised Continuing Appropriations Resolution to review DOE's execution of the LGP and to report its findings to the House and Senate Committees on Appropriations. This is the sixth time we have reported on this program.[2] We have raised concerns in our prior work about the limitations of the portion of the program conducted under section 1703 in attracting financially viable projects representing the full range of targeted technologies.[3] In addition, we previously reported, among other things, that the LGP treated applicants inconsistently and recommended that DOE treat applicants consistently or clearly establish the conditions that would warrant disparate treatment.[4] Because of questions regarding inconsistent treatment of applicants and DOE's review process that we raised in the 2010 report, our objectives for this report were to determine (1) the status of the applications to the LGP's nine solicitations and (2) the extent to which the LGP has adhered to its process for reviewing applications for loans that the LGP has committed to or closed.

To determine the status of applications to the LGP's nine solicitations, we reviewed DOE and LGP documents on the establishment and operation of the program and analyzed the LGP's available data on the applications received and their current status. Because the LGP did not maintain consolidated information on application status, it had to assemble data from various sources for all of the applications as of September 30, 2011. To assist in this effort, we tailored a data request to collect data on the status of all 460 applications to the program in consultation with agency officials. These data were to provide a current snapshot of the program by solicitation and allow analysis of various characteristics. LGP staff familiar with each solicitation completed the spreadsheets, and these spreadsheets were reviewed by managers before they were forwarded to GAO. We assessed the reliability of the data the LGP provided by reviewing it, comparing it to other sources and following up with the agency to clarify questions and inconsistencies, and obtain missing data. Once the data were all collected, we found them to be sufficiently reliable for our purposes. This process enabled us to develop up-to-date programwide information on the status of applications. The LGP staff updated its March 2011 applicant status data as of July 29, 2011, and we obtained additional data on the conditional commitments and closings made by the September 30, 2011, expiration of the section 1705 authority for loan guarantees with a credit subsidy. In cases where multiple applications were submitted for a single project, we considered each to be a single application for purposes of this report. In addition, we met with the LGP's management and staff from each of the divisions involved with the review process. To determine the extent to which the LGP has adhered to its process for reviewing applications for loans that it has committed to or closed, we identified the key steps in the review process by analyzing the laws, regulations, policies, guidance, and solicitations for the program.

We verified these key steps in interviews with LGP officials. We identified the 13 applications that had received conditional commitments or had closed by December 31, 2010.[5] We then requested documentation from the LGP of the key review steps it conducted for selected applications. We initially requested this documentation for a nonprobability sample of 6 applications representing a range of solicitations and project types.[6] We also collected more limited information on the 7 remaining applications to which DOE had conditionally committed to issue a loan guarantee by the end of calendar year 2010. For these 7, we reviewed certain key steps for which we found differences from the LGP's established process during our review of the initial 6 applications. We did not evaluate the quality of the LGP's analyses supporting the completion of these steps. The applications we reviewed were processed by the LGP under the policies and procedures that were in place through September 30, 2011.[7]

We also interviewed seven private lenders with experience financing energy sector projects to gain insights on the comparability of the LGP and private sector review processes. A more detailed discussion of our objectives, scope, and methodology is presented in appendix I.

We conducted this performance audit from September 2010 to February 2012 in accordance with generally accepted government auditing standards. Those standards require that we plan and perform the audit to obtain sufficient, appropriate evidence to provide a reasonable basis for our findings and conclusions based on our audit objectives. We believe that the evidence obtained provides a reasonable basis for our findings and conclusions based on our audit objectives.

BACKGROUND

DOE's LGP was designed to address the fundamental impediment for investors that stems from the high risks of clean energy projects, including technology risk—the risk that the new technology will not perform as expected—and execution risk—the risk that the borrower will not perform as expected. Companies can face obstacles in securing enough affordable financing to survive the "valley of death" between developing innovative technologies and commercializing them. Because the risks that lenders must assume to support new technologies can put private financing out of reach, companies may not be able to commercialize innovative technologies without the federal government's financial support. According to the DOE loan program's Executive Director, DOE loan guarantees lower the cost of capital for projects using innovative energy technologies, making them more competitive with conventional technologies and thus more attractive to lenders and equity investors. Moreover, according to the DOE loan programs Executive Director, the program takes advantage of DOE's expertise in analyzing the technical aspects of proposed projects, which can be difficult for private sector lenders without that expertise.

Until February 2009, the LGP was working exclusively under section 1703 of the Energy Policy Act of 2005, which authorized loan guarantees for new or innovative energy technologies that had not yet been commercialized. Congress had authorized DOE to guarantee approximately $34 billion in section 1703 loans by fiscal year 2009, after accounting for rescissions, but it did not appropriate funds to pay the "credit subsidy costs" of these guarantees. For section 1703 loan guarantees, each applicant was to pay the credit subsidy cost of its own project. These costs are defined as the estimated long-term cost, in net present value terms, over the entire period the loans are outstanding to cover interest subsidies, defaults, and delinquencies (not including administrative costs). Under the Federal Credit Reform Act of 1990, the credit subsidy cost for any guaranteed loan must be provided prior to a loan guarantee commitment.

In past reports, we found several issues with the LGP's implementation of section 1703. For example, in our July 2008 report, we stated that risks inherent to the program make it difficult for DOE to estimate credit subsidy costs it charges to borrowers.[8] If DOE underestimates these costs, taxpayers will ultimately bear the costs of defaults or other shortfalls not covered by the borrowers' payments into a cost-subsidy pool that is to cover section 1703's program-wide costs of default. In addition, we reported that, to the extent that certain types of projects or technologies are more likely than others to have fees that are too high to remain economically viable, the projects that do accept guarantees may be more heavily weighted toward lower-risk technologies and may not represent the full range of technologies targeted by the section 1703 program.

In February 2009, the Recovery Act amended the Energy Policy Act of 2005, authorizing the LGP to guarantee loans under section 1705. This section also provided $2.5 billion to pay applicants' credit subsidy costs.[9] This credit subsidy funding was available only to projects that began construction by September 30, 2011, among other requirements.[10] DOE estimated that the funding would be sufficient to provide about $18 billion in guarantees under section 1705. Section 1705 authorized guarantees for commercial energy projects that employ renewable energy systems, electric power transmission systems, or leading-edge biofuels that meet certain criteria. Some of these are the same types of projects eligible under section 1703, which authorizes guarantees only for projects that use new or significantly improved technologies.[11] Consequently, many projects that had applied under section 1703 became eligible to have their credit subsidy costs paid under section 1705.

Because authority for the section 1705 loan guarantees expired on September 30, 2011, section 1703 is now the only remaining authority for the LGP. In April 2011, Congress appropriated $170 million to pay credit subsidy costs for section 1703 projects. Previously, these costs were to be paid exclusively by the applicants and were not federally funded. Congress also authorized DOE to extend eligibility under section 1703 to certain projects that had applied under section 1705 but did not receive a loan guarantee prior to the September 30, 2011, deadline.[12]

DOE has issued nine calls for applications to the LGP. Each of these nine "solicitations" has specified the energy technologies it targets and provided criteria for the LGP to determine project eligibility and the likelihood of applicants repaying their loans (see table 1).

Table 1. DOE Solicitations for Applications to the LGP

Name of solicitation	Date issued or updated	Description of eligible energy technology
Mixed 06	8/8/06	All technologies except for nuclear facilities and oil refineries.
Nuclear Front-End	6/30/08	Facilities for new uranium enrichment capacity and distribution.
Nuclear Power	6/30/08	Nuclear power facilities.
Energy efficiency and renewable energy or EERE 08	6/30/08	Innovative energy efficiency, renewable energy, and advanced energy transmission and distribution technologies.
Fossil	9/22/08	Coal-based power generation and industrial gasification facilities that incorporate carbon capture and sequestration or other beneficial uses of carbon and for advanced coal gasification facilities.
Energy efficiency and renewable energy or EERE 09	7/29/09	Innovating energy efficiency, renewable energy, and advanced energy transmission and distribution technologies.
Transmission	7/29/09	Electric power transmission infrastructure investment projects.
Financial Institution Partnership Program (FIPP)	10/7/09	Renewable energy generation projects using commercial technology.
Manufacturing	8/10/10	Manufacture of renewable energy systems and components using commercial technology.

Source: DOE.

To help ensure that that these criteria were applied consistently and that each selected project provided a reasonable prospect of repayment, in March 2009, the LGP issued a credit policies and procedures manual for the program, outlining its policies and procedures for reviewing loan guarantee applications. As shown in figure 1, this review process is divided into three stages: intake, due diligence, and "conditional commitment to closing." We use the term "review process" to refer to the entire process.

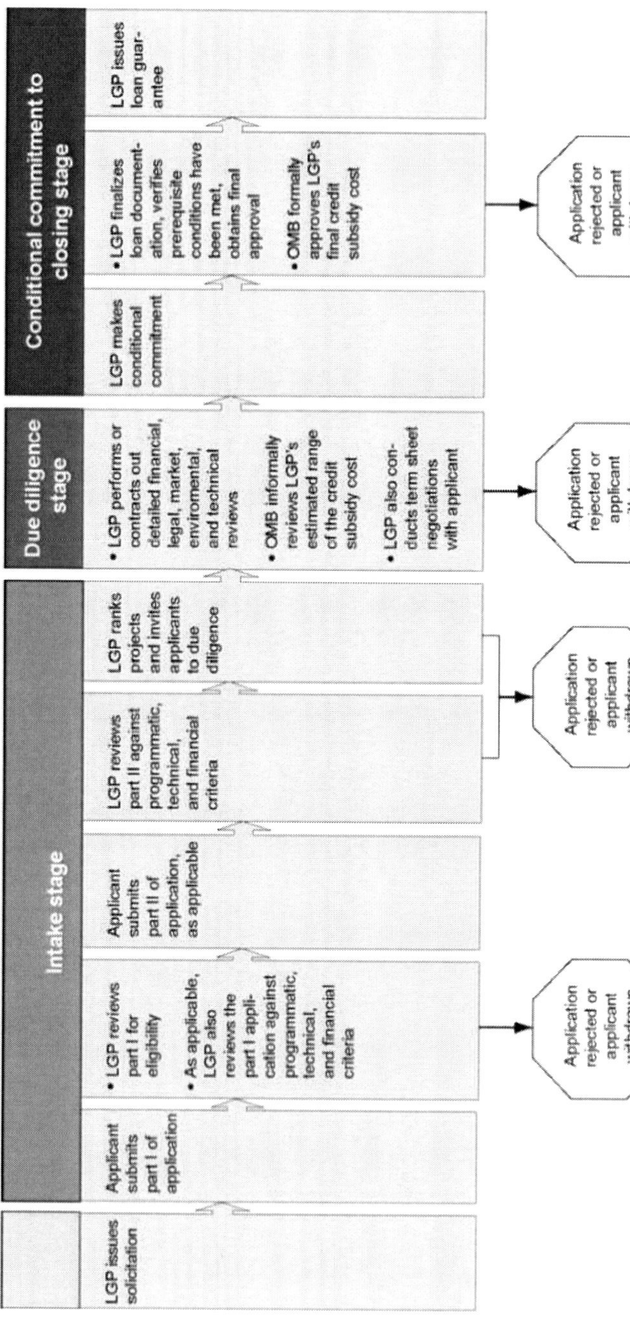

Source: GAO presentation of DOE data.

Figure 1. Overview of LGP Review Process for Applications through the Intake, Due Diligence, and Conditional Commitment to Closing Stages.

applicants can withdraw at any point during the review process. Once these steps have been completed, the LGP "closes" the loan guarantee and, subject to the terms and conditions of the loan guarantee agreement, begins to disburse funds to the project. For further detail on the review process, see appendix III.

DOE HAS MADE $15.1 BILLION IN LOAN GUARANTEES BUT DOES NOT MAINTAIN CONSOLIDATED DATA ON STATUS OF APPLICATIONS

For 460 applications to the LGP from its nine solicitations, DOE has made $15.1 billion in loan guarantees and conditionally committed to an additional $15 billion, representing $30 billion of the $34 billion in loan guarantees authorized for the LGP.[14] However, when we requested data from the LGP on the status of the applications to its nine solicitations, the LGP did not have consolidated data readily available but had to assemble them from various sources.

DOE Has Made $15.1 Billion in Loan Guarantees and Committed to Another $15 Billion

As of September 30, 2011, the LGP had received 460 applications and made (closed) $15.1 billion in loan guarantees in response to 30 applications (7 percent of all applications), all under section 1705. It had not closed any guarantees under section 1703.

In addition, the LGP had conditionally committed another $15 billion for 10 more applications (2 percent of all applications)—4 under section 1705 and 6 under section 1703. The closed loan guarantees obligated $1.9 billion of the $2.5 billion in credit subsidy appropriations funded by the Recovery Act for section 1705, leaving $600 million of the funds unused before the program expired.

For section 1703 credit subsidy costs, the $170 million that Congress appropriated in April 2011 to pay such costs is available, but it may not cover all such costs because the legislation makes the funds available only for renewable energy or efficient end-use energy technologies.[15] Applicants whose projects' credit subsidy costs are not covered by the appropriation must pay their own credit subsidy costs.

During the intake stage, the LGP assesses applications in a two-par process for most applicants. In part I, the LGP considers a project's eligibility based on the requirements in the solicitation and relevant laws and regulations. Nuclear solicitation applications are also evaluated against programmatic, technical, and financial criteria during the part I review. Based on the LGP's eligibility determination during part I review, qualifying applicants are invited to submit a part II application. Generally, LGP evaluates this application against programmatic, technical, and financial criteria to form a basis for ranking applications within each solicitation.[13] Based on these initial rankings, the LGP selects certain applications for the due diligence stage. During due diligence, the LGP performs a detailed examination of the project's financial, technical, legal, and other qualifications to ensure that the LGP has identified and mitigated any risks that might affect the applicant's ability to repay the loan guarantee. Key to identifying risks during due diligence are required reports by independent consultants on the technical and legal aspects of the project and others, such as marketing reports, that the LGP uses when needed. The LGP also negotiates the terms of the loan guarantee with the applicant during due diligence.

The proposed loan guarantee transaction is then submitted for review and/or approval by the following entities:

- DOE's Credit Committee, consisting of senior executive service DOE officials, most of whom are not part of the LGP.
- DOE's Credit Review Board (CRB), which consists of senior-level officials such as the deputy and undersecretaries of Energy.
- The Office of Management and Budget (OMB), which reviews the LGP's estimated credit subsidy range for each transaction.
- Department of the Treasury.
- The Secretary of Energy, who has final approval authority.

Following the Secretary's approval, the LGP offers the applicant a "conditional commitment" for a loan guarantee. If the applicant signs and returns the conditional commitment offer with the required fee, the offer becomes a conditional commitment, contingent on the applicant meeting conditions prior to closing. During the conditional commitment to closing stage, LGP officials and outside counsel prepare the final financing documents and ensure that the applicant has met all conditions required for closing, and the LGP obtains formal approval of the final credit subsidy cost from OMB. Prior to closing, applications may be rejected by the LGP. Similarly,

Table 2. Number of Applications, Median and Total Loan Guarantees Requested, Total Conditionally Committed Loan Guarantees, and Total Closed Loan Guarantees, by Solicitation, through September 30, 2011

Dollars in millions					
Solicitation, issue date	Number of applications	Median loan guarantee requested	Total loan guarantee requested	Total conditionally committed loan guarantee	Total closed loan guarantee[a]
Mixed 06, 8/8/06	140	$60[b]	$31,018[b]	$72[c]	$1,203
Energy Efficiency and Renewable Energy 08, 6/30/08[c]	68	163	21,265	261[c]	3,381
Nuclear Front-End, 6/30/08	2	2,000	4,000	2,000[c]	0
Nuclear Power, 6/30/08	19	6,969	117,363	8,326[c]	0
Fossil, 9/22/08	8	2,072	17,145	0	0
Energy Efficiency and Renewable Energy 09, 7/29/09	168	150	52,915	2,105	5,601
Transmission, 7/29/09	12	660	11,586	0	343
Financial Institution Partnership Program, 10/7/09	37	146	11,057	2,274	4,516
Manufacturing, 8/10/10	6	98	1,022	0	0
Total	460	$141[d]	$267,372	$15,038[e]	$15,044

Source: GAO analysis of DOE data provided as of July 29, 2011, and updated for new commitments or closings, as of September 30, 2011.

Note: Totals may not add due to rounding.

[a]Fifteen of these guarantees went to projects that applied under section 1703 but were later deemed eligible for and received funding under section 1705

[b]The median and total loan guarantee amounts reflect the reported loan amounts for 134 of the 140 Mixed 06 applications because DOE said 6 applicants did not specify the amount of their loan guarantee request.

[c]This row includes four applications that LGP does not consider to be official submissions since the applicants did not pay the application fee. However, we included these applications in our analysis because the LGP included them in the application data they provided to us, and these applications demonstrate the level of interest in the solicitation.

[d] This amount is the median loan guarantee amount requested across all solicitations. The minimum loan guarantee requested for all applications was $0, and the maximum loan guarantee requested was $12 billion, both for nuclear power projects.

[e] Of the $15 billion in committed loan guarantees, applications under the section 1703 authority to these solicitations account for $10.4 billion or 71 percent. See appendix II for a list of these committed loan guarantees.

To date, credit subsidy costs for loan guarantees that DOE has closed have, on average, been about 12.5 percent of the guaranteed loan amounts. The median loan guarantee requested for all applications was $141 million. Applications for nuclear power projects requested significantly larger loan amounts—a median of $7 billion—and requested the largest total dollar amount by type of technology—$117 billion.[16] Applications for energy efficiency and renewable energy solicitations requested the second-largest dollar amount—$74 billion. Table 2 provides further details on the applications by solicitation and the resulting closed loan guarantees and conditional commitments. Appendix II provides further details on the individual committed and closed loan guarantees.

For all 460 LGP applications submitted, figure 2 shows the total loan guarantee amounts requested by type of energy technology.

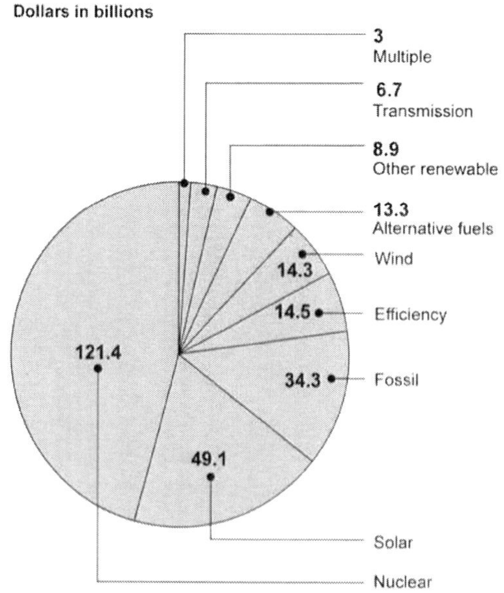

Source: GAO analysis of DOE data describing the energy technology of each loan guarantee application, as of July 29, 2011.

Note: For this analysis, we used simplified energy technology categories based on DOE's data. The figure omits one application for which the LGP did not report the type of energy technology employed by the proposed project or the amount requested for the project. It also omits requests that DOE listed as using "other" energy technology, which totaled about 0.01 percent of the amount requested.

Figure 2. Amount of Loan Guarantees Requested in 460 Applications by Energy Technology Category, as of July 29, 2011.

Table 3 provides an overview, as of September 30, 2011, of the status of the 460 loan guarantee applications that the LGP received in response to its nine solicitations.

Table 3. Number (and Percentage) of Applications in Each Review Stage, by Solicitation, as of September 30, 2011

Solicitation, issue date	Number of applications	Number (percentage) of applications by stage				Withdrawn	Rejected
		Intake	Due diligence	Conditional commitment	Guarantees made (closed)		
Mixed 06, 8/8/06	140	0(0)	3(2)	1(1)	4(3)	8(6)	124(89)
EERE 08, 6/30/08[a]	68	0(0)	7(10)	2(3)	8(12)	6(9)	45(66)
Nuclear Front End, 6/30/08	2	0(0)	1(50)	1(50)	0(0)	0(0)	0(0)
Nuclear Power, 6/30/08	19	5(26)	4(21)	3(16)	0(0)	7(37)	0(0)
Fossil, 9/22/08	8	0(0)	4(50)	0(0)	0(0)	3(38)	1(13)
EERE 09, 7/29/09	168	12(7)	21(13)	1(1)	10(6)	59(35)	65(39)
Transmission, 7/29/09	12	2(17)	0(0)	0(0)	1(8)	9(75)	0(0)
Financial Institution Partnership Program (FIPP), 10/7/09	37	0(0)	7(19)	2(5)	7(19)	18(49)	3(8)
Manufacturing, 8/10/10	6	0(0)	0(0)	0(0)	0(0)	3(50)	3(50)
Total[b]	460	19(4)	47(10)	10(2)[c]	30(7)[d]	113(25)	241(52)

Source: GAO analysis of DOE data provided as of July 29, 2011 and updated for new commitments or closings as of September 30, 2011.

[a] This row includes four applications that the LGP does not consider to be official submissions since the applicants did not pay the application fee. However, we included these applications in our analysis because the LGP included them in the application data they provided to us, and these applications demonstrate the level of interest in the solicitation.

[b] Percentage totals may not add to 100 due to rounding.

[c] Four of these conditional commitments are under section 1705, and six are under section 1703. Many of the section 1703 applications have been in process since 2008 or before. See appendix II, tables 7 and 8.

[d] All of these closed loan guarantees are under section 1705. See appendix II, table 9.

Of the 460 applications, 66 were still in various stages of the approval process (intake and due diligence), 40 had received conditional commitment or were closed, and 354 had been withdrawn or rejected. DOE documents list a wide range of reasons for application withdrawals, including inability to submit application material in a timely manner, inability to secure feedstock, project faced many hurdles, applicant did not pursue project, and applicant switched to another program.

Solicitations that primarily targeted efficiency and renewable energy received the most applications, while those targeting nuclear front-end technologies (for the beginning of the nuclear fuel cycle), manufacturing, and fossil fuels received the fewest. The rejection rate was highest for applications submitted for two of the earlier solicitations and much lower for DOE's FIPP,[17] a more recent solicitation involving applications sponsored by private financial institutions.

Since we began our review, two of the borrowers with closed loan guarantees have declared bankruptcy—Solyndra, Inc., with a $535 million loan guarantee for manufacturing cylindrical solar cells, and Beacon Power Corporation, with a $43 million loan guarantee for an energy storage technology. The elapsed time for LGP to process loan applications generally decreased over the course of the program, according to LGP data. LGP officials noted that the elapsed time between review stages includes the time the LGP waited for the applicants to prepare required documents for each stage. The process was longest for applications to the earlier solicitations, issued solely under section 1703, from start to closing.[18] The review process was shorter for applications under the four more recent solicitations, issued after the passage of section 1705. For example, the first solicitation, known as Mixed 06, had the longest overall time frames from intake to closing—a median of 1,442 days—and the FIPP solicitation had the shortest time frames—a median of 422 days.[19] Applications to the FIPP solicitation had the shortest elapsed time because this program was carried out in conjunction with private lenders, who conducted their own reviews before submitting loan applications to the LGP.[20] Table 4 shows the median number of days elapsed during each review stage, by solicitation, as of September 30, 2011.

From September 4, 2009, to July 29, 2011—a period of nearly 2 years— the LGP closed $5.8 billion in loan guarantees for 13 applications under section 1705. In the last few months before the authority for section 1705 loan guarantees expired, the LGP accelerated its closings of section 1705 applications that had reached the conditional commitment stage.

Table 4. Number of Applications and Median Number of Days Elapsed During Each Review Stage and Overall, by Solicitation

Solicitation, issue date	Number of applications	Intake stage (parts I & II)	Due diligence stage	Conditional commitment to closing stage	Overall: start of intake to closing date	Rejected applications:start of intake to rejection date
Number of applications completing this stage		114	43	30	30	241
Mixed 06, 8/8/06	140	722	430	284	1,442	280
EERE 08, 6/30/08 [a]	68	90	338	177	696	168
Nuclear Front-End, 6/30/08	2	230	401	b	b	b
Nuclear Power, 6/30/08	19	219	294	b	b	b
Fossil, 9/22/08	8	199	b	b	b	201
EERE 09, 7/29/09	168	194	309	114	668	70
Transmission, 7/29/09	12	178	222	115	515	b
FIPP, 10/7/09	37	73	228	92	422	94
Manufacturing, 8/10/10	6	b	b	b	b	78
Median number of days for allapplications		*184[c]*	*294[d]*	*127[e]*	*660[f]*	*277[g]*

Source: GAO analysis of DOE data.

Notes: The number of days elapsed represents the median for the review period of those applications that proceeded to the next review stage. We believe the median is a better representation of the data for this table because it reduces the effect of some outliers that skew the data. The calculations were for the status and elapsed days of the 460 applications as of September 30, 2011. The elapsed time between review stages includes the time the LGP waited for the applicant to prepare required documents for each stage.

[a] This row includes four applications that the LGP does not consider to be official submissions since the applicants did not pay the application fee. However, we included these applications in our analysis because the LGP included them in the application data they provided to us, and these applications demonstrate the level of interest in the solicitation. Additionally, the average elapsed time for intake review of applications to the EERE 08 solicitation is lower because applications for certain types of projects were not required to follow a two-part application process. [b] No applications completed this stage for this solicitation. [c] The minimum number of days elapsed for all applications for intake was 33 and the maximum number of days elapsed was 740. [d] The minimum number of days elapsed for all applications for due diligence was 50 and the maximum number of days elapsed was 930. [e] The minimum number of days elapsed for all applications for conditional commitment to closing was 41 and the maximum number of days elapsed was 407. [f] The minimum number of days elapsed for all applications from intake to closing was 287 and the maximum number of days elapsed was 1,731. [g] The minimum number of days elapsed for all applications from intake to rejection was 9 and the maximum number of days elapsed was 1,046.

Thus, over the last 2 months before the authority for section 1705 expired, the LPG closed an additional $9.3 billion in loan guarantees for 17 applications under section 1705.[21] The program did not use about $600 million of the $2.5 billion that Congress appropriated to pay credit subsidy costs before the section 1705 authority expired, and these funds were no longer available for use by LGP.

The LGP Does Not Maintain Consolidated Information on Application Status

When we requested data from the LGP on the identity of applicants, status, and key dates for review of all the applications to its nine solicitations, the LGP did not have consolidated information on application status readily available. Instead, it had to assemble these data from various sources.

To respond to our initial data request, LGP staff provided information from the following five sources:

- "Origination portfolio" spreadsheets, which contain information for applications that are in the due diligence stage of the review process. These spreadsheets contain identifying information, the solicitation applied under, commitment or closing status, type of technology, overall cost, proposed or closed loan amount, and expected or actual approval dates. Information in these spreadsheets is limited. For example, they do not contain dates that the applicant completed each stage and do not have information on applications that have been rejected or withdrawn.

- "Tear sheet" summaries for each application, which give current status and basic facts about the project and its technology, cost, finances, and strengths and weaknesses. Tear sheets are updated periodically, or as needed, but LGP officials could not easily consolidate them because they were kept in word processing software that does not have analysis or summarization capabilities.

- "Application trackers," which are spreadsheets that give basic descriptive information and status of applications for some solicitations. LGP staff said they were maintained for most, but not all, solicitations.

- "Project Tracking Information" documents showing graphic presentations of application status summaries, loan guarantee amounts requested, technology type, planned processing dates, and procurement schedules for technical reports. These documents were updated manually through December 20, 2010.

- "Credit subsidy forecasts," which are documents that track the actual or projected credit subsidy costs of the section 1705 projects in various stages of the review process and the cumulative utilization of credit subsidy funding.

LGP staff needed over 3 months to assemble the data and fully resolve all the errors and omissions we identified. LGP staff also made further changes to some of these data when we presented our analysis of the data to the LGP in October 2011.[22]

According to LGP officials in 2010, the program had not maintained up-to-date and consolidated documents and data. An LGP official said at the time that LGP considered it more important to process loan guarantee applications than to update records. Because it took months to assemble the information required for our review, it is also clear that the LGP could not be conducting timely oversight of the program.

Federal regulations require that records be kept to facilitate an effective and accurate audit and performance evaluation. These regulations—along with guidance from the Department of the Treasury and OMB—provide that maintaining adequate and proper records of agency activities is essential to oversight of the management of public resources.[23]

In addition, under federal internal control standards, federal agencies are to employ control activities, such as accurately and promptly recording transactions and events to maintain their relevance and value to management on controlling operations and making decisions.[24]

Under these standards, managers are to compare actual program performance to planned or expected results and analyze significant differences. Managers cannot readily conduct such analysis of the LGP if the agency does not maintain consolidated information on applications to the program and their status.

Moreover, the fact that it took the LGP 3 months to aggregate data on the status of applications for us suggests that its managers have not had readily accessible and up-to-date information and have not been doing such analysis on an ongoing basis. This is not consistent with one of the fundamental concepts of internal control, in which such control is not a single event but a series of actions and activities that occur throughout an entity's operations and on an ongoing basis.

Thus, providing managers with access to aggregated, updated data could facilitate more efficient management of the LGP. Furthermore, without consolidated data about applicants, LGP actions, and application status, LGP staff may not be able to identify weaknesses, if any, in the program's application review process and approval procedures. For example, consolidated data on application status would provide a comprehensive snapshot of which steps of the review process are taking longer than expected and may need to be addressed.

If program data were consolidated in an electronic tracking system, program managers could quickly access information important to managing the LGP, such as the current amount of credit subsidy obligated, as well as whether the agency is consistently complying with certain procedural requirements under its policies and regulations that govern the program. In addition, the program cannot quickly respond to requests for information about the program as a whole from Congress or program auditors.

In March 2011, the LGP acknowledged the need for such a system. According to the March 2011 LGP summary of its proposed data management project, as the number of applications, volume of data and records, and number of employees increased, the existing method for storing and organizing program data and documents had become inadequate, and needed to be replaced. In October 2011, LGP officials stated that while the LGP has not maintained a consolidated application tracking database across all solicitations, the program has started to develop a more comprehensive business management system that includes a records management system called "iPortal" that also could be used to track the status of applications. Officials did not provide a timetable for using iPortal to track the status of applications but said that work is under way on it. However, until iPortal or some other system can track applications' status, the LGP staff cannot be assured that consolidated information on application status necessary to better manage the program will be available.

THE LGP DID NOT ALWAYS ADHERE TO ITS REVIEW PROCESS, WHICH MAY POSE RISKS AND RESULT IN INCONSISTENT TREATMENT

We identified 43 key steps in the LGP's guidance establishing its review process for assessing and approving loan guarantee applications. The LGP followed most of its established review process, but the LGP's actual process differed from this established process at least once on 11 of the 13 applications we reviewed, in part because the process was outdated. In some cases, LGP did not perform applicable review steps and in other cases we could not determine whether the LGP had completed review steps.

Furthermore, we identified more than 80 instances of deficiencies in documentation of the LGP's reviews of the 13 applications, such as missing signatures or dates.

It is too early to evaluate the impact of the specific differences we identified on achieving program goals, but we and the DOE Inspector General have reported that omitting or poorly documenting review steps may pose increased financial risk to the taxpayer and result in inconsistent treatment of applications.

The LGP Did Not Consistently Follow Its Established Review Process, in Part Because the Process Was Outdated

We identified 43 key steps in the LGP credit policies and procedures manual and its other guidance that establish the LGP's review process for assessing and approving loan guarantee applications.

Not all 43 steps are necessary for every application, since the LGP's guidance lets officials tailor aspects of the review process on an ad hoc basis to reflect the specific needs of the solicitation.

For example, under the EERE 08 solicitation, the LGP required two parts of intake review for applications involving large projects that integrate multiple types of technologies, but it required only one part for small projects. Furthermore, according to LGP officials, they have changed the review process over time to improve efficiency and transparency, so the number of relevant steps also depends on when the LGP started reviewing a given application.

LGP guidance recognizes the need for such flexibility and maintains that program standards and internal control need to be applied transparently and uniformly to protect the financial interests of the government. For more information on the key steps we identified, see appendix III.

According to private lenders we contacted who finance energy projects, the LGP's established review process is generally as stringent as or more stringent than those lenders' own due diligence processes. For example, like the LGP, private lenders evaluate a project's proposed expenses and income in detail to determine whether it will generate sufficient funds to support its debt payments. In addition, private lenders and the LGP both rely on third-party expertise to evaluate the technical, legal, and marketing risks that might affect the payments. Lenders who were not participating in the LGP generally agreed that the LGP's process, if followed, should provide reasonable management of risk. Some lenders that sponsored applications under the FIPP solicitation said that the LGP's review process was more rigorous than their own. They said this level of rigor was not warranted for the FIPP solicitation because it covered commercial technology, which is inherently less risky than the innovative technologies covered by other solicitations.

Some private lenders we spoke with also noted that financing an innovative energy project involves a certain amount of risk that cannot be eliminated, and one lender said that a failure rate of 2 or 3 percent is common, even for the most experienced loan officers. However, we found that the LGP did not always follow the review process in its guidance. The LGP completed most of the applicable review steps for the 6 applications that we reviewed in full, but its actual process differed from the established process at least once on 5 of the 6 applications we reviewed.

We also conducted a more limited examination of 7 additional applications, in which we examined the steps where the actual process differed from the established process for the first 6 applications.

We again found that the LGP's actual process differed from its established process at least once on 6 of the 7 applications.

Table 4 summarizes review steps for which we either identified differences or could not determine whether the LGP completed a particular review step across all 13 applications.

The 13 applications we reviewed represent all of the applications that had reached conditional commitment or closing, as of December 31, 2010, excluding 3 applications that had applied under the earliest solicitation, since the LGP's review process was substantially different for these 3 applications.[25]

For the 13 applications we examined, we found 19 differences between the actual reviews the LGP conducted and the applicable review process steps established in LGP guidance.

In most of these instances, according to LGP officials, the LGP did not perform an applicable review step because it had made changes intended to improve the process but had not updated the program's credit policies and procedures manual or other guidance governing the review process.

Table 5. LGP's Adherence to Its Review Process for 13 Applications with Closed or Conditionally Committed Loan Guarantees, by Review Stage and Step

Review stage	Review step description	Number of applications examined for review step	Not applicable	Applicable but not performed	Could not determine if step was performed	Completed
Intake	Part I technical review	6	2	0	1	3
	Solicitation-specific ranking process for EERE 08 applicants	13	5	0	8	0
	Obtain CRB approval prior to due diligence	13	3	6	0	4
Due diligence	Review of applicant's management (e.g., background check, credit check, Internal Revenue Service check)	6	0	2	1	3
	Obtain final independent engineering report prior to conditional commitment	13	0	6	0	7
	Obtain final independent marketing report prior to conditional commitment	13	3	1	1	8
	Complete OMB review of the LGP credit subsidy cost estimate	13	0	3	7	3
Conditional commitment to closing	Collect a full fee from an applicant at conditional commitment	13	0	1	0	12
Total			13	19	18	40

Source: GAO analysis of LGP documentation supporting its application reviews.

Note: These differences represent our review of LGP documents for all 43 key review steps for six applications, and a targeted review of 9 steps for seven applications.

The following describes the 19 differences we identified, along with the LGP's explanations:

- In six cases, the LGP did not obtain CRB approval prior to due diligence, contrary to the March 2009 version of its credit policies and procedures manual. This version states that CRB approval is an important internal check to ensure only the most promising projects proceed to due diligence. LGP officials explained that this step was not necessary for these applications because the CRB had verbally delegated to the LGP its authority to approve applications before these projects proceeded to due diligence. However, LGP documents indicate that CRB delegated approval authority after these projects had proceeded to due diligence.[26] According to an LGP official, the delegation of authority was not retroactive.

- In seven cases, the LGP did not obtain final due diligence reports from independent consultants prior to conditional commitment, as required by its credit policies and procedures manual. Through their reporting, these independent third parties provide key input to the LGP's loan underwriting and credit subsidy analyses in technical, legal, and other areas such as marketing, as necessary. LGP officials said that it was a preferable practice to proceed to conditional commitment with drafts of these reports and obtain a final report just prior to closing. They said this practice helps the LGP reduce financial risk, since it allows the LGP to base its decision to close the loan guarantee on final reports rather than reports completed 1 to several months earlier. An LGP official explained that this part of the review process had evolved to meet the program's needs, but that these changes were not yet reflected in the manual. However, the LGP does not appear to have implemented this change consistently. Specifically, over the course of several months in 2009 and 2010, the LGP alternated between the old and the new process concerning final due diligence reports from independent consultants. In commenting on a draft of this report, LGP officials said that in all cases they received final independent consultant reports before the closing of the loan guarantees. Because the LGP's policies and procedures manual at the time required final reports at the conditional commitment stage, we reviewed the reports available at conditional commitment and did not review whether LGP received final reports before closing.

- In three cases, the LGP conditionally committed to a loan guarantee before OMB had completed its informal review of the LGP's credit subsidy cost estimate. According to the credit policies and procedures manual, OMB should be notified each time the LGP estimates the credit subsidy cost range, and informal discussions between OMB and LGP should ensue about the LGP estimate. This cost is to be paid by the borrower for all section 1703 projects to date and by the federal government for section 1705 projects. LGP officials explained that, in two of these cases, the LGP had provided OMB with their credit subsidy estimates, but that OMB had not completed its review because there were unresolved issues with the LGP estimates. LGP officials did not provide an explanation for the third case. Contrary to the manual, LGP officials said that OMB's informal review of the credit subsidy estimates for these applications was not a necessary prerequisite to conditional commitment because the actual credit subsidy cost is calculated just prior to closing and is formally approved by OMB. Furthermore, under section 1705, the government rather than the borrower, was to pay credit subsidy costs. Accordingly, the LGP used these credit subsidy estimates for internal planning purposes rather than for calculating a fee to the applicant. In contrast, the LGP completed OMB's informal review prior to conditionally committing to at least three of the other loan guarantees we reviewed—including one section 1705 project—and thus the LGP did not perform this step consistently across all projects. In its October 2011 update of its credit policies and procedures manual, the LGP retained the requirement that OMB review the LGP's credit subsidy cost estimate prior to conditional commitment. Further, the updated guidance added that formal discussions with OMB may be required each time OMB reviews LGP's credit subsidy cost estimate and should result with their approval. In two cases, the LGP did not complete its required background check for project participants. The documents provided indicate that LGP did not determine whether the applicants had any delinquent federal debt prior to conditional commitment. In one of these cases, LGP officials said that the delinquent federal debt check was completed after conditional commitment. In the other case, the documents indicate that the sponsor did not provide a statement on delinquent debt, and LGP officials confirmed that LGP did not perform the delinquent debt check prior to conditional commitment.

- In one case, the LGP did not collect the full fee from an applicant at conditional commitment as required by the EERE 08 solicitation. According to a LGP official, the LGP changed its policy to require 20 percent of this fee at conditional commitment instead of the full fee specified in the solicitation, in response to applicant feedback. This official said the policy change was documented in the EERE 09 solicitation, which was published on July 29, 2009. However, this particular application moved to conditional commitment on July 10, 2009, prior to the formal policy change.

As outlined in these cases, the LGP departed from its established procedures because, in part, the procedures had not been updated to reflect all current review practices. The version of the manual in use at the time of GAO's review was dated March 5, 2009, even though the manual states that it was meant to be updated at least on an annual basis and more frequently if needed. The LGP issued its first update of its credit policies and procedures manual on October 6, 2011,[27] even though the 2009 manual states that it was meant to be updated at least annually and more frequently if needed. We reviewed the revised manual and found that the revisions addressed many of the differences that we identified between the LGP's established and actual review processes.

The revised manual also states that LGP analyses should be properly documented and stored in the new LGP electronic records management system. However, the revised guidance applies to loan guarantee applications processed after October 6, 2011, but not to the 13 applications we reviewed or to any of the 30 loan guarantees the LGP has closed to date.

The LGP Did Not Always Fully Document Review Steps

In addition to the differences between the actual and established review processes, in another 18 cases, we could not determine whether the LGP had performed a given review step. In some of these cases, the documentation did not demonstrate that the LGP had applied the required criteria. In other cases, the documentation the LGP provided did not show that the step had been performed. The following discusses these cases:

- In one case, we could not determine whether LGP guidance calls for separate part I and part II technical reviews for a nuclear front-end application or allows for a combined part I and part II technical review. The LGP performed a combined part I and part II technical review.

- In eight cases, we could not determine the extent to which the LGP applied the required criteria for ranking applications to the EERE 08 solicitation. The LGP's guidance for this solicitation requires this step to identify "early mover" projects for expedited due diligence. The LGP expedited four such applications but the documentation neither demonstrated how the LGP used the required criteria to select applications to expedite nor why other applications were not selected.

- In one case, we could not determine whether the LGP completed its required background check for project participants. The documents provided indicated there were unresolved questions involving one participant's involvement in a $17 billion bankruptcy and another's pending civil suit.

- In one case, we could not determine whether the LGP had received a draft or final marketing report prior to conditional commitment in accordance with its guidance. The LGP provided a copy of the report prepared before closing but did not provide reports prepared before conditional commitment.

- In seven cases, LGP either did not provide documents supporting OMB's completion of its informal review of the LGP's estimated credit subsidy range before conditional commitment, or the documentation the LGP provided was inconclusive.

We also found 82 additional documentation deficiencies in the 13 applications we reviewed. For example, in some cases, there were no dates or authors on the LGP documents. The documentation deficiencies make it difficult to determine, for example, whether steps occurred in the correct order or were executed by the appropriate official. The review stage with the fewest documentation deficiencies was conditional commitment to closing, when 1 of the 82 deficiencies occurred. Table 6 shows the instances of deficient documentation that we identified.

During our review, the LGP did not have a central paper or electronic file containing all the documents supporting the key review steps we identified as being part of the review process. Instead, these documents were stored separately by various LGP staff and contractors in paper files and various electronic storage media. As a result, the documents were neither readily available for us to examine, nor could the LGP provide us with complete documentation in a timely manner. For example, we requested documents supporting the LGP's review for six applicants in January 2011.

**Table 6. Documentation Deficiencies Identified During
13 Application Reviews**

Review phase	Missing author	Missing title or other identification	Missing final version or a signature	Missing date	Missing data or analysis	Inconsistent with other project documents
Intake	15	6	2	17	8	1
Due diligence	12	0	6	6	5	3
Conditional commitment to closing	0	0	0	0	0	1
Total	27	6	8	23	13	5

Source: GAO analysis of LGP documentation supporting its application reviews.

Note: These deficiencies represent our review of LGP documents for all 43 review steps for six applications and a targeted review of 9 steps for seven applications.

For one of the applications, we did not receive any of the requested documents supporting the LGP's intake application reviews until April 2011. Furthermore, for some of the review steps, we did not receive documents responsive to our request until November 2011 and, as we discussed earlier, in 18 cases we did not receive sufficient documentation to determine whether the LGP performed a given review step. Federal regulations and guidance from Treasury and OMB provide that maintaining adequate and proper records of agency activities is essential to accountability in the management of public resources and the protection of the legal and financial rights of the government and the public.[28] Furthermore, under the federal standards for internal control, agencies are to clearly document internal control, and the documentation is to be readily available for examination in paper or electronic form.

Moreover, the standards state that all documentation and records should be properly managed and maintained.[29]

As stated above, the LGP recognized the need for a recordkeeping system to properly manage and maintain documentation supporting project reviews. In March 2011, the LGP adopted a new records management system called "iPortal" to electronically store documents related to each loan application and issued guidance for using this system. As of November 1, 2011, LGP officials told us that the system was populated with data or records relevant to conditionally committed and closed loan guarantees and that they plan to fully populate it with documentation of the remaining applications in a few months. The LGP was able to provide us with some additional documents from its new system in response to an early draft of this report, but the LGP did not provide additional documentation sufficient to respond to all of the issues we identified. Accordingly, other oversight efforts may encounter similar problems with documentation despite the new system.

Differences between the Actual and Established Processes and Incomplete Documentation May Pose Risks

It is too early in the loan guarantees' terms to assess whether skipping or poorly documenting review steps will result in problems with the guarantees or the program. However, we and the DOE Inspector General have reported that omitting or poorly documenting review steps may lead to a risk of default or other serious consequences. Skipping or poorly documenting steps of the process during intake can lead to several problems. First, it reduces the LGP's assurance that it has treated applications consistently and equitably. This, in turn, raises the risk that the LGP will not select the projects most likely to meet its goals, which include deploying new energy technologies and ensuring a reasonable prospect of repayment. In July 2010, we reported that the inconsistent treatment of applicants to the LGP could also undermine public confidence in the legitimacy of the LGP's decisions. Furthermore, DOE's Inspector General reported in March 2011 that incomplete records may impede the LGP's ability to ensure consistency in the administration of the program, make informed decisions, and provide information to Congress, OMB, and other oversight bodies.[30] The Inspector General also stated that, in the event of legal action related to an application, poor documentation of the LGP's decisions may hurt its ability to prove that it applied its procedures consistently and treated applicants equitably. Moreover, incomplete records may leave DOE open to criticism that it exposed taxpayers to unacceptable financial risks.

Differences between the actual and established review processes that occur during or after due diligence may also lead to serious consequences. These stages of the review process were established to help the LGP identify and mitigate risks. Omitting or poorly documenting its decisions during these stages may affect the LGP's ability to fully assess and communicate the technical, financial, and other risks associated with projects. This could lead the program to issue guarantees to projects that pose an unacceptable risk of default. Complete and thorough documentation of decisions would further enable DOE to monitor the loan guarantees as projects are developed and implemented. Furthermore, without consistent documentation, the LGP may not be able to fully measure its performance and identify any weaknesses in its implementation of internal procedures.

CONCLUSION

Through the over $30 billion in loan guarantees and loan guarantee commitments for new and commercial energy technologies that DOE has made to date, the agency has set in motion a substantial federal effort to promote energy technology innovation and create jobs. DOE has also demonstrated its ability to make section 1705 of the program functional by closing on 30 loan guarantees. It has also improved the speed at which it was able to move section 1705 applications through its review process. To date, DOE has committed to six loan guarantees under section 1703 of the program, but it has not closed any section 1703 loan guarantees or otherwise demonstrated that the program is fully functional. Many of the section 1703 applications have been in process since 2008 or before. As DOE continues to implement section 1703 of the LGP, it is even more important that it fully implement a consolidated system for overseeing the application review process and that LGP adhere to its review process and document decisions made under updated policies and procedures. It is noteworthy that the process LGP developed for performing due diligence on loan guarantee applications may equal or exceed those used by private lenders to assess and mitigate project risks. However, DOE does not have a consolidated system for documenting and tracking its progress in reviewing applications fully implemented at this time. As a result, DOE may not readily access the information needed to manage the program effectively and to help ensure accountability for federal resources. Proper recordkeeping and documentation of program actions is essential to effective program management. The absence of such documentation may have prevented LGP managers, DOE, and Congress from having access to the timely and accurate information on applications necessary to manage the program, mitigate risk, report progress, and measure program performance.

DOE began to implement a new records management system in 2011, and LGP staff stated that the new system will enable them to determine the status of loan guarantee applications and to document review decisions. However, the LGP has neither fully populated the system with data or records on all applications it has received nor its decisions on them. Nor has DOE committed to a timetable to complete the implementation of the new records management system. Until the system has been fully implemented, it is unclear whether the system will enable the LGP to both track applications and adequately document its review decisions.

In addition, DOE did not always follow its own process for reviewing applications and documenting its analysis and decisions, potentially increasing the taxpayer's exposure to financial risk from an applicant's default. DOE has not promptly updated its credit policies and procedures manual to reflect its changes in program practices, which has resulted in inconsistent application of those policies and procedures.

It also has not completely documented its analysis and decisions made during reviews, which may undermine applicants' and the public's confidence in the legitimacy of its decisions. Furthermore, the absence of adequate documentation may make it difficult for DOE to defend its decisions on loan guarantees as sound and fair if it is questioned about the justification for and equity of those decisions.

DOE has recently updated its credit policies and procedures manual, which, if followed and kept up to date, should help the agency address this issue.

RECOMMENDATIONS FOR EXECUTIVE ACTION

To better ensure that LGP managers, DOE, and Congress have access to timely and accurate information on applications and reviews necessary to manage the program effectively and to mitigate risks, we recommend that the Secretary of Energy direct the Executive Director of the Loan Programs Office to take the following three actions:

- Commit to a timetable to fully implement a consolidated system that enables the tracking of the status of applications and that measures overall program performance.
- Ensure that the new records management system contains documents supporting past decisions, as well as those in the future.

- Regularly update the LGP's credit policies and procedures manual to reflect current program practices to help ensure consistent treatment for applications to the program.

AGENCY COMMENTS AND OUR EVALUATION

We provided a copy of our draft report to DOE for review and comment. In written comments signed by the Acting Executive Director of the Loan Programs Office, it was unclear whether DOE generally agreed with our recommendations. The Acting Executive Director stated subsequently to the comment letter that DOE disagreed with the first recommendation and agreed with second and third recommendations. In its written comments, DOE also provided technical and editorial comments, which were incorporated as appropriate.

Concerning our first recommendation that LGP commit to a timetable to fully implement a consolidated system that enables the tracking of the status of applications and that measures overall program performance, in its written comments, DOE states that the LGP believes that it is important that our report distinguish between application tracking and records management. We believe we have adequately distinguished the need for application tracking and management of documentation.

These are addressed in separate sections of our report and in separate recommendations. DOE also states that LGP has placed a high priority on records management and is currently implementing a consolidated state-of-the-art records management system.

In the statement subsequent to DOE's written comments, the Acting Executive Director stated the office did not agree to a hard timetable for implementing our first recommendation. As stated in the report draft, under federal internal control standards, agencies are to employ control activities, such as accurately and promptly recording transactions and events to maintain their relevance and value to management on controlling operations and making decisions. Because LGP had to manually assemble the application status information we needed for this review, and because this process took over 3 months to accomplish, we continue to believe DOE should develop a consolidated system that enables the tracking of the status of applications and that measures overall program performance. This type of information will help LGP better manage the program and respond to requests for information from Congress, auditors, or other interested parties.

Concerning our second recommendation that LGP ensure that its new records management system contains documents supporting past decisions as well as those in the future, subsequent to DOE's written comments, the Acting Executive Director stated that DOE agreed.

Concerning our third recommendation that LGP regularly update the credit policies and procedures manual to reflect current program practices, subsequent to DOE's written comments, the Acting Executive Director stated that DOE agreed.

Frank Rusco
Director, Natural Resources and Environment

APPENDIX I: OBJECTIVES, SCOPE, AND METHODOLOGY

This appendix details the methods we used to examine the Department of Energy's (DOE) Loan Guarantee Program (LGP). We have reported four times and testified three times on this program, including two previous reports in response to the mandate in the 2007 Revised Continuing Appropriations Resolution to review DOE's execution of the LGP and to report our findings to the House and Senate Committees on Appropriations. (See Related GAO Products.) Because of questions regarding inconsistent treatment of applications raised by the most recent report in this mandated series,[1] this report, also in response to the mandate, assesses (1) the status of the applications to the LGP's nine solicitations and (2) the extent to which the LGP has adhered to its process for reviewing applications for loans that the LGP has committed to or closed.

To gather information on the program, we met with the LGP's management and staff from each of the program's divisions involved with the LGP's review of loan guarantee applications from intake to closing. In general, we reviewed the laws, regulations, policies and procedures governing the program and pertinent agency documents, such as solicitations announcing loan guarantee opportunities. We reviewed prior GAO and DOE Inspector General reports performed under or related to our mandate to audit the LGP. In addition, we gathered agency data and documents on the loan guarantee applications in process, those that had received a DOE commitment, and those that had been closed.

To determine the status of the applications to all nine of the solicitations for our first objective, we explored the LGP's available sources to see what data the program had compiled on the applications received and their current status in the review process. Because the LGP did not have comprehensive or complete application status data, we tailored a data request to collect data on the status of all 460 applications to the program. In consultation with agency officials, we prepared a data collection form requesting basic information on the identity, authority, amount requested, status, key milestone dates, and type of energy technology for all of the applications to date. These data were to provide a current snapshot of the program by solicitation and allow analysis of various characteristics. To ease the data collection burden, we populated the spreadsheets for each solicitation with the limited data from available sources. LGP staff or contractors familiar with each solicitation completed the spreadsheets, and these spreadsheets were reviewed by managers before they were forwarded to GAO. We assessed the reliability of the data the LGP provided by reviewing these data, comparing them to other sources, and following up repeatedly with the agency to clarify questions and inconsistencies, and obtain missing data. This process enabled us to develop up-to-date program-wide information on the status of applications. This process resulted in data that were complete enough to describe the status of the program. Once we collected these data, we found them to be sufficiently reliable for our purposes. The LGP updated its March 2011 applicant status data as of July 29, 2011, and we obtained additional data on the conditional commitments and closings made by the September 30, 2011, expiration of the section 1705 authority for loan guarantees with a credit subsidy. To maintain consistency between the application status data initially provided by the LGP and later data updates, we use the terms application and project interchangeably, although in some cases multiple applications were submitted for a single project.

To assess the LGP's execution of its review process for our second objective, we first analyzed the law, regulations, policies, procedures, and published solicitations for the program and interviewed agency staff to identify the criteria and the key review process steps for loan guarantees, as well as the documents that supported the process. We provided a list of the key review steps we identified to LGP officials, and incorporated their feedback as appropriate. Based on the key review steps and supporting documentation identified by LGP staff, we developed a data collection instrument to analyze LGP documents and determine whether the LGP followed its review process for the applications reviewed. Since the LGP's review process varied across solicitations, we tailored the data collection instrument to meet the needs of the individual solicitations. We then selected a nonprobability sample of 6 applications from the 13 that had received conditional commitments from DOE or had progressed to closing by December 31, 2010, and had not applied under the Mixed 2006 solicitation, since the LGP's review process was substantially different for this solicitation and not directly comparable to later solicitations.[2] We requested documentation for these 6 applications representing a range of solicitations and project types. We selected our initial sample to represent each of the five solicitations where applications had reached conditional commitment and different LGP investment officers to reduce the burden on LGP staff. We requested the documents supporting the LGP's review process from intake to closing and examined them to determine whether the applicable review steps were carried out. While we examined whether the applicable review steps were carried out, we did not examine the content of the documents and the quality of work supporting them. Where the documents were not clear about completion of the process, showed potential differences from the review process, or raised questions, we followed up with program officials to obtain an explanation and, as applicable, documentation supporting the explanation. On key questions where we identified differences from the review process for the initial sample of 6, we conducted a targeted review of documents for the 7 remaining applications that had reached conditional commitment or closed prior to December 31, 2010, excluding Mixed 2006 applicants. The six loan guarantee application files reviewed in full and the seven files reviewed in part were a nongeneralizable sample of applications.

To identify the initial universe of private lenders with experience financing energy projects, we reviewed the list of financial institutions that had submitted applications to the LGP under the Financial Institution Partnership Program (FIPP) solicitation. We used these firms as a starting point because of their knowledge about DOE's program and processes. To identify financial institutions involved in energy sector project finance outside of FIPP, we searched or contacted industry associations, industry conferences, and other industry groups in the same energy sectors that LGP solicitations to date have targeted. We interviewed seven private lenders identified through this process using a set of standard questions and the outline of the DOE's review process to gain insights on its comparability to the review process for underwriting loans in the private sector.

We conducted this performance audit from September 2010 to February 2012 in accordance with generally accepted government auditing standards. Those standards require that we plan and perform the audit to obtain sufficient, appropriate evidence to provide a reasonable basis for our findings and conclusions based on our audit objectives. We believe that the evidence obtained provides a reasonable basis for our findings and conclusions based on our audit objectives.

APPENDIX II: TABLES OF LOAN GUARANTEES CONDITIONALLY COMMITTED OR CLOSED

The following tables provide basic details on the loan guarantee applications that received a conditional commitment by September 30, 2011, or had proceeded to closing by that date. Table 7 lists applications under section 1703 with conditional commitments. Table 8 lists section 1705-eligible applications with conditional commitments that did not reach closing by the expiration of the section 1705 authority on September 30, 2011. Table 9 lists the section 1705 applications with conditional commitments that reached closing by the expiration of the section of the 1705 authority on September 30, 2011.

Table 7. Section 1703 Applications Reaching Conditional Commitment as of September 30, 2011, by Solicitation

Dollars in millions					
Solicitation	Sponsor	Name	Technology	Date conditional commitment offered	Guarantee amount
Mixed, 8/8/06	SAGE Electrochromics, LLC	SAGE Electrochromics	Energy Efficiency	3/5/2010	$72
EERE 08, 6/30/08	ADA-ES, Inc.	Red River	Energy Efficiency	12/8/2009	245
Nuclear Front-End, 6/30/08	AREVA NC, Inc.	Eagle Rock Enrichment Facility	Nuclear Front-End	5/20/2010	2,000
Nuclear Power, 6/30/08	Georgia Power Company	Vogtle 3&4	Nuclear Generation	2/16/2010	3,460
Nuclear Power, 6/30/08	MEAG	Vogtle 3&4	Nuclear Generation	2/16/2010	1,809
Nuclear Power, 6/30/08	Oglethorpe Power Corp.	Vogtle 3&4	Nuclear Generation	2/16/2010	3,057
Total					$10,643

Source: GAO analysis of DOE data.

Table 8. Section 1705-Eligible Applications Reaching Conditional Commitment as of September 30, 2011, by Solicitation

Dollars in millions					
Solicitation	*Sponsor*	*Name*	*Technology*	*Date conditional commitment offered*	*Guarantee amount*
EERE 08, 6/30/08	Nordic Windpower, Ltd.	Nordic Project	Wind Manufacturing	7/2/2009	$16
EERE 09, 7/29/09	Solar Millennium, LLC	Blythe Solar Power Project Plant	Solar Generation	4/18/2011	2,105
FIPP, 10/7/09	First Solar	Topaz (CA)	Solar Generation	6/30/2011	1,930
FIPP, 10/7/09	Multiple	SolarStrong (USA)	Solar Generation	9/8/2011	344
Total					*$4,395*

Source: GAO analysis of DOE data.

Table 9. Section 1705-Eligible Applications Reaching Closing as of September 30, 2011, By Solicitation

Dollars in millions						
Solicitation	*Sponsor*	*Name*	*Technology*	*Date conditional commitment offered*	*Date closed*	*Guarantee amount*
Mixed, 8/8/06	Beacon Power Corp.	Beacon Power	Transmission	7/2/2009	8/6/2010	$43
Mixed, 8/8/06	BrightSource Energy, Inc.	Ivanpah 1	Solar Generation	2/22/2010	4/5/2011	520
Mixed, 8/8/06	POET, LLC	Project LIBERTY	Biomass	7/7/2011	9/23/2011	105
Mixed, 8/8/06	Solyndra, Inc.	Solyndra Fab 2, LLC	Solar Manufacturing	3/20/2009	9/4/2009	535
EERE 08, 6/30/08	Abengoa Solar, Inc.	Solana Project	Solar Generation	7/2/2010	12/20/2010	1,446
EERE 08, 6/30/08	Abound Solar, Inc.	Abound Solar Manufacturing, LLC	Solar Manufacturing	7/2/2010	12/9/2010	400
EERE 08, 6/30/08	AES Energy Storage, LLC	Project Dyno	Transmission	7/30/2010	12/22/2010	17

Table 9. (Continued)

Dollars in millions						
Solicitation	*Sponsor*	*Name*	*Technology*	*Date conditional commitment offered*	*Date closed*	*Guarantee amount*
EERE 08, 6/30/08	Bright-Souce Energy, Inc.	Ivanpah 2	Solar Generation	2/22/2010	4/5/2011	551
EERE 08, 6/30/08	Bright-Souce Energy, Inc.	Ivanpah 3	Solar Generation	2/22/2010	4/5/2011	556
EERE 08, 6/30/08	First Wind Energy, LLC	Kahuku Wind Power	Wind Generation	2/18/2010	7/26/2010	117
EERE 08, 6/30/08	SoloPower Inc.	SoloPower Manufacturing Facility	Solar Manufacturing	2/17/2011	8/19/2011	197
EERE 08, 6/30/08	U.S. Geothermal, Inc.	Neal Hot Springs	Geothermal	6/9/2010	2/23/2011	97
EERE 09, 7/29/09	1366 Technologies, Inc.	Project Eagle	Solar Maufacturing	6/17/2011	9/8/2011	150
EERE 09, 7/29/09	Abengoa Bioenergy U.S. Holding	Abengoa Bioenergy Biomass of Kansas	Biomass	8/19/2011	9/29/2011	132
EERE 09, 7/29/09	Abengoa Solar, Inc.	Mojave Solar Project	Solar Generation	6/14/2011	9/23/2011	1,202
EERE 09, 7/29/09	Cogentrix Solar Services, LLC	Alamosa Solar Generating Project	Solar Generation	5/10/2011	9/9/2011	91
EERE 09, 7/29/09	Nextlight Renewable Energy, LLC	Antelope Valley Solar Ranch 1	Solar Generation	6/30/2011	9/30/2011	646
EERE 09, 7/29/09	Nextlight Renewable Energy, LLC	Agua Caliente	Solar Generation	1/20/2011	8/5/2011	967
EERE 09, 7/29/09	Sempra Generation	Mesquite Solar Energy	Solar Generation	6/15/2011	9/28/2011	337
EERE 09, 7/29/09	Solar Reserve, LLC	Tonopah Project	Solar Generation	5/19/2011	9/28/2011	737
EERE 09, 7/29/09	Sunpower Corp.	California Valley Solar Ranch	Solar Generation	4/12/2011	9/30/2011	1,237
EERE 09, 7/29/09	Yale University	Record Hill Wind	Wind Generation	3/3/2011	8/15/2011	102
Transmission 7/29/09	LS Power AssociatesLP	Southwest Intertie Project (SWIP) - South	Transmission	10/19/2010	2/11/2011	343

Dollars in millions						
Solicitation	*Sponsor*	*Name*	*Technology*	*Date conditional commitment offered*	*Date closed*	*Guarantee amount*
FIPP, 10/7/09	Caithness Energy, LLC	Shepherds Flat (OR)	Wind Generation	10/8/2010	12/16/2010	1,051
FIPP, 10/7/09	First Solar	Desert Sun (CA)	Solar Generation	6/30/2011	9/30/2011	1,169
FIPP, 10/7/09	Nevada Geothermal Power Company	Blue Mountain (NV)	Geothermal	6/15/2010	9/3/2010	79
FIPP, 10/7/09	NextEra	Genesis Solar (CA)	Solar Generation	6/14/2011	8/26/2011	682
FIPP, 10/7/09	Noble Environmental Power, LLC	Noble Granite (NH)	Wind Generation	6/21/2011	9/23/2011	135
FIPP, 10/7/09	Ormat Nevada, Inc.	Ormat (NV)	Geothermal	6/9/2011	9/23/2011	280
FIPP, 10/7/09	Prosun Solar Development Company, LLC	Project Amp (USA)	Solar Generation	6/22/2011	9/30/2011	1,120
Total						$15,044

Source: GAO analysis of DOE data.

Appendix III: Key Tasks in the LGP's Review and Approval Process for Loan Guarantee Applications

Table 10 provides basic details about key review tasks in LGP's process for reviewing and approving loan guarantee applications, as identified from our review of relevant laws, regulations, LGP guidance, published solicitations and interviews with LGP officials. These tasks formed the basis for our examination of LGP files to determine if LGP followed its review process for each of the 13 applications that had received conditional commitments from DOE or had progressed to closing by December 31, 2010, and had not applied under the Mixed 2006 solicitation.[1] Accordingly, the tasks listed below reflect

LGP's review process for the applications we reviewed and do not reflect LGP's review process for applicants to the Mixed 2006 solicitation, which was substantially different and not directly comparable to later solicitations. Additionally, since we found minor variations in LGP's review process across the solicitations, we have noted below which tasks are only applicable under certain solicitations. If no exceptions are listed, then the particular task is applicable across all the relevant solicitations.

Table 10. Key Review and Approval Tasks for Loan Guarantee Applications, by Review Stage

Review stage and task	Description
Intake	
1. Collect part I application fee.	The first of three fees that LGP collects during the review process. LGP is required by
	its authorizing legislation to charge and collect sufficient fees to cover the program's administrative costs.
2. Perform part I completeness check.	LGP reviews applications using a solicitation-specific checklist to document that the
	application package is complete.
3. Perform innovation review (EERE 08 applicants).	LGP reviews applications to determine if the proposed project uses an innovative
	energy technology, as required by the program's authorizing legislation. For later
	solicitations, this review was incorporated into the LGP's technical review.
4. Perform part I technical review (2008 Nuclear Power and Nuclear Front-End) or commercial review (FIPP).	LGP analyzes the project's eligibility and responsiveness to statutory and program requirements, such as the project's • technical relevance against DOE requirements, • technical approach and work plan, and • environmental and technological benefits.
5. Perform part I financial review (2008 Nuclear Power and Nuclear Front-End).	LGP analyzes • creditworthiness elements such as sponsor/management capabilities, financial/business plans, and market factors; and • programmatic elements such as (a) construction and start-up factors and (b) legal, regulatory, and permitting factors.
6. Perform emissions review or lifecycle analysis (EERE 08 and EERE 09 applicants).	Loan guarantee applications under the EERE 08 and EERE 09 solicitations must pass an emissions analysis to meet the authorizing law's greenhouse gas emissions goals.
7. Perform review for solicitation-specific eligibility requirements. [a]	Depending on the solicitation, loan guarantee applications must meet certain solicitation-specific eligibility requirements, related to • certain project types,

Review stage and task	Description
	certain technology categories, and • construction commencement requirements for section 1705 projects
7. Perform review for solicitation-specific eligibility requirements. [a]	Depending on the solicitation, loan guarantee applications must meet certain solicitation-specific eligibility requirements, related to • certain project types, • certain technology categories, and • construction commencement requirements for section 1705 projects.
8.a. Rank projects to identify "Early Movers" (EERE 08 only).	LGP identifies the projects that present the fewest obstacles in moving forward to begin the technical and financial review process first. The ranking factors are related to • level of environmental review required under the National Environmental Policy Act of 1970, • financial structure, • readiness to proceed, and • offtake agreements if applicable (an agreement to buy all or a substantial part of the output of an energy project).
8.b. Rank Projects and Identify Project Strengths and Weaknesses as part of the part I review (2008 Nuclear Power and Nuclear Front-End).	The 2008 Nuclear Power and Front-End solicitations call for an early ranking of projects. The ranking factors are related to • the prospect of repayment, • strength of the project and sponsor, and • regulatory status.
9. Notify applicants of intent to proceed/invite part II submissions (part II submissions exclude certain EERE 08 projects).	For solicitations with a one-part intake process, applicants are notified of LGP's intent to proceed with its review. For solicitations with a two-part intake process, applicants are notified they have qualified under part I and are invited to submit application materials for part II.
10. Collect part II application fee. [b]	The second of three fees that LGP collects during the review process. LGP is required by the authorizing legislation to charge and collect sufficient fees to cover the program's administrative costs.
11. Perform part II completeness check.	LGP reviews applications using a solicitation-specific checklist to document that the part II application package is complete.
12. Perform part II technical review (excludes FIPP).	LGP analyzes • the project's technical relevance against DOE requirements, • track record and experience of applicant, • project work plan, and • environmental benefits of project.
13. Perform part II financial review (excludes FIPP).	LGP analyze • creditworthiness elements such as sponsor/management capabilities, financial/business plans, and market factors; and • programmatic elements such as (a) construction and start-up factors and (b) legal, regulatory, and permitting factors.

Table 10. (Continued)

Review stage and task	Description
14. Perform an environmental critique and synopsis.	LGP may prepare a publicly available environmental critique and synopsis to document the consideration given to environmental factors and record that the relevant environmental consequence of each alternative has been considered in its evaluation and selection process.
15. Application screening/ranking sessions for finalization of merit review scores for selections to due diligence.	To focus limited loan guarantee funds on the best applicants, LGP evaluates and competitively ranks all applications within each solicitation's cohort. This ranking is the basis for LGP's decision to invite applicants to due diligence.
16. DOE's Credit Review Board (CRB) approves projects recommended for due diligence by LGP (only projects proceeding to due diligence prior to 6/25/09).	DOE's CRB reviews LGP's recommendations of projects for due diligence and provides approval. The CRB delegated this authority to LGP on June 25, 2009, and this task was phased out for applications proceeding to due diligence following this decision.
17. Notify applicant of LGP's decision to proceed into due diligence (excludes FIPP).	After clearing requirements of parts I and II, the applicants are notified that they will proceed into due diligence.
Due diligence	
18. Evaluate financing plan and assess financial viability.	To evaluate the project in detail, LGP will • thoroughly review the uses and sources of funds; • analyze adequacy, leverage, timing of funding; • review terms/rights of funding source; • assess the adequacy of proposed contingency/reserve funding; • determine compliance with program requirements from the law, final regulations, and the solicitation; • assess the project's financial viability, with an emphasis on the applicant's ability to repay the guaranteed portion of loan; and • evaluate assumptions underlying projected revenues/expenses/likelihood technical performance will be achieved.
19. Perform a review of applicant's management.	LGP performs certain checks (e.g., background check, credit check, IRS check) to evaluate the key players for the loan guarantee applicant.
20. Evaluate project risks and identify risk mitigants.	To evaluate the project's risks and potential mitigants, LGP will • identify, assess, and estimate the impact of risks associated with the project; • determine the types and magnitude of the risks associated with the project; • determine the proper risk allocation among the parties; and • determine the extent to which risks have been mitigated.

Review stage and task	Description
21. Perform a financial model analysis and stress-test.	To evaluate the project's financial model, LGP will • verify the applicant's calculations for its financial model, and • quantify the impacts of risks by stress-testing the applicant's and LGP's financial models for changes in assumptions.
22. Assess strengths and weaknesses of project participants.	LGP will examine the sponsor's investment to date and financial/managerial capability to implement the project as proposed, including • the project sponsor's track record in project development and the technology used in the application, • the project sponsor's financial strength and resources, • the strategic value of the project to the sponsor, and • the experience of the project's management team.
23. Assess whether an environmental assessment, environmental impact statement, or categorical exclusion applies. (For FIPP projects, this assessment step occurs during intake.)	As required by the National Environmental Policy Act of 1970 (NEPA), LGP reviews the project and determines which environmental review process is necessary.
24. Prepare Environmental Assessment, Environmental Impact Statement or Categorical Exclusion.	Based on LGP's analysis under task 23, LGP prepares the appropriate documents, which include a description of any significant findings under other applicable environmental laws.
25. Identify significant findings under other applicable environmental laws.	
26. Receive independent engineering/technical consultant report.	To determine the technical efficacy of the project, LGP or an independent engineering firm, will thoroughly review the applicant's independent engineering report, including consideration of factors such as environmental impact and infrastructure requirements. This review also provides input for the risks and mitigants section of the credit paper.
27. Receive independent legal analysis.	To review the project's legal structure, LGP or an external firm will • analyze draft legal agreements among project participants, • analyze intellectual property rights of participants in the project to use the proposed technology, and • provide input for the risks and mitigants section of the credit paper.
28. Receive independent marketing consultant report (as applicable).	As necessary, LGP will consult with external marketing advisors to assess the project's market and off-take risk as part of the underwriting and credit analysis process. This assessment should be supported by data, examples, and/or research that substantiate the score assigned for each attribute.

Table 10. (Continued)

Review stage and task	Description
29. Negotiate term sheet.	Based on its due diligence analysis and input from any external advisors, LGP prepares a term sheet and negotiates its provisions with the applicant.
30. Calculate expected recovery rate.	LGP calculates the percentage of value the agency can expect to recover in the event of default.
31. Prepare a credit approval package.	LGP assembles key documents describing the proposed loan guarantee agreement and project for internal review. These include • the credit paper providing an overview of the project and its attributes, • available third-party input, • draft term sheet, • internal risk rating matrix, • recovery rate notching matrix, • compliance checklist, and • presentation summarizing the transaction for internal and external review.
32. Credit committee reviews and approves the credit approval package.	LGP management internal review and approval step.
Review stage and task	Description
33. Office of Management and Budget (OMB) reviews LGP's credit subsidy estimate.	OMB reviews LGP's calculation of the estimated credit subsidy cost range for the project and provides informal approval. The credit subsidy cost is based on a formula designed to determine the net present value of the estimated cost to the federal government of guaranteeing the loan.
34. LGP consults with U.S. Treasury regarding the commitment of Federal Financing Bank funds.	The Department of the Treasury reviews the transaction.
35. DOE's CRB approves projects recommended by LGP for conditional commitment.	DOE leadership review and approval step.
Conditional commitment to closing	
36. DOE offers applicant conditional commitment for a loan guarantee and applicant accepts.	DOE conditionally commits to issuing a loan guarantee agreement dependent upon whether the conditions precedent laid out in the term sheet are met. Upon accepting the offer, the applicant pays all or a portion of the second fee, depending on the solicitation.
37. LGP prepares and negotiates definitive financing documentation	LGP and external counsel prepare and negotiate the final financing terms and loan guarantee agreement.

Review stage and task	Description
38. LGP receives final credit rating from a rating agency via the applicant.	The applicant obtains and provides final credit rating to LGP.
39. LGP legal team circulates an action memo to all relevant parties for concurrence and the Secretary's signature.	Internal review and approval step that includes a crosswalk between the key terms at the time of conditional commitment and the final closing terms, including any material adverse differences.
40. OMB formally approves the final credit subsidy cost.	OMB review and key decision step.
41. Outside counsel confirms that all conditions precedent to the loan guarantee agreement have been satisfied.	LGP asks outside counsel to verify that the applicant has met all of the terms agreed to at conditional commitment as preconditions for LGP's approval of the final loan guarantee agreement.
42. DOE and applicant execute loan agreement, and DOE issues guarantee.	The final loan guarantee documents are executed at closing and the loan is considered closed once the agreements have been executed.
43. First funds disbursement.	At the time of or shortly after the loan guarantee's closing, the Federal Financing Bank, or other lender, disburses the first payment of funds to the loan guarantee recipient.

Sources: GAO analysis of DOE guidance, published solicitations, and relevant regulations.

[a] According to LGP officials, this step is a component of the innovation and other eligibility reviews rather than a separate step. However, we included it as a separate step in our list of key review tasks since it was an important aspect of the process.

[b] As applicable, for solicitations where LGP established a two part application process for some or all applicants (excludes stand-alone or manufacturing projects that applied under EERE 08)

End Notes

[1] The amount of authority does not include approximately $20 billion that expired when authority for a portion of the program expired on September 30, 2011.

[2] See our list of related products on the LGP at the end of this report.

[3] GAO, Department of Energy: New Loan Guarantee Program Should Complete Activities Necessary for Effective and Accountable Program Management, GAO-08-750 (Washington, D.C.: July 7, 2008).

[4] GAO-08-750; GAO, Department of Energy: Further Actions Are Needed to Improve DOE's Ability to Evaluate and Implement the Loan Guarantee Program, GAO-10-627 (Washington, D.C.: July 12, 2010).

[5] Three additional applications had either reached conditional commitment or closed during this period. We excluded these applications from our review because the LGP's review process

for these applications was substantially different. A conditional commitment is a commitment by DOE to issue a loan guarantee if the applicant satisfies specific requirements. The Secretary of Energy has the discretion to cancel a conditional commitment at any time for any reason prior to the issuance of a loan guarantee.

[6] Because this was a nonprobability sample, we cannot generalize what we found to all applications, but we chose these applications to include a variety of project types and for different solicitations.

[7] Department of Energy, Title XVII Of The Energy Policy Act Of 2005 Loan Guarantee Program, Credit Policies and Procedures (Washington, D.C., Mar. 5, 2009).

[8] GAO-08-750.

[9] Pub. L. No. 111-5, Div. A, Title IV (Feb. 17, 2009). Congress originally appropriated nearly $6 billion to pay the credit subsidy costs of projects supported under section 1705, with the limitation that funding to pay the credit subsidy costs of leading-edge biofuel projects eligible under this section would not exceed $500 million. Congress later authorized the President to transfer up to $2 billion of the nearly $6 billion to expand the "Cash for Clunkers" program. Pub. L. No. 111-47 (Aug. 7, 2009). The $2 billion was transferred to the Department of Transportation, leaving nearly $4 billion to cover credit subsidy costs of projects supported under section 1705. On August 10, 2010, Pub. L. No. 111-226 rescinded an additional $1.5 billion from the loan guarantee appropriation to pay for education-related jobs, Medicaid and other initiatives, further reducing funding available to $2.5 billion.

[10] Other requirements include that the workers employed on the project, including contractors or subcontractors, will be paid wages not less than prevailing on similar work in the locality in accordance with the Davis-Bacon Act. The act limited loan guarantees under section 1705 to the following categories of projects: (1) renewable energy systems, including incremental hydropower, that generate electricity or thermal energy, and facilities that manufacture related components; (2) electric power transmission systems, including upgrading and reconductoring projects; and (3) leading-edge biofuel projects that will use technologies performing at the pilot or demonstration scale that the Secretary determines are likely to become commercial technologies and will produce transportation fuels that substantially reduce life-cycle greenhouse gas emissions compared with other transportation fuels.

[11] New or significantly improved technology means a technology concerned with the production, consumption, or transportation of energy and that is not a commercial technology, and that has either: (1) only recently been developed, discovered, or learned; or (2) involves or constitutes one or more meaningful and important improvements in productivity or value, in comparison to commercial technologies in use in the United States at the time the term sheet is issued.

[12] Department of Defense and Full-Year Continuing Appropriations Act, 2011, Pub. L No. 112-10.

[13] Under the FIPP solicitation, applicants must apply to a private "lead lender," which initially evaluates the proposed loan guarantee for credit approval and decides whether to apply to DOE for the loan guarantee.

[14] The amount of authority does not include approximately $20 billion that expired when authority for a portion of the program expired on September 30, 2011.

[15] The legislation also made some section 1705 projects submitted to DOE by Feb. 24, 2011, eligible for these funds, but nuclear projects are not included among eligible projects.

[16] The minimum loan guarantee requested for all applications was $0, and the maximum loan guarantee requested was $12 billion, both for nuclear power projects. The $0 loan guarantee

request was for one portion of a jointly sponsored nuclear power project. The joint sponsor of the nuclear power project requested approximately $8 billion in loan guarantees.

[17] FIPP refers to the federal loan guarantees for commercial technology renewable energy generation projects under the DOE LGP solicitation number DE-FOA-0000166, dated October 7, 2009. This solicitation is unique because it is the only one inviting private lenders to share due diligence activities for identifying and mitigating risk and finance a portion of total project costs.

[18] Some closed loan guarantees went to projects that applied under section 1703 but were later eligible for and received funding under section 1705.

[19] The minimum number of days elapsed from intake to closing was 287 days, and the maximum number of days from intake to closing was 1,731 days.

[20] The LGP's lenders to date have been the U.S. Treasury's Federal Financing Bank and the private lenders under the FIPP solicitation who brought applicants to the LGP and who must risk at least 20 percent of the total loans for the applicants' project.

[21] This effort was preceded by DOE's announcement on May 10, 2011, that the LGP would focus on 18 applications that officials believed were most likely to meet all the requirements for closing and begin construction prior to the September 30, 2011, expiration date.

[22] Errors and omissions included missing or incorrect dates associated with an applicant's progression to the next stage of LGP review; incorrect status (e.g., application listed as both withdrawn and rejected); inconsistent entries related to the loan guarantee amount requested by the applicant; and no status given.

[23] 10 C.F.R. § 609.10(f)(1); U.S. Department of the Treasury, Managing Federal Receivables A Guide for Managing Loans and Administrative Debt, Financial Management Service (Washington, D.C.: 2005); and OMB Circular A-130.

[24] GAO, Standards for Internal Control in the Federal Government, GAO/AIMD-00-21.3.1 (Washington, D.C.: November 1999).

[25] The three excluded applications were from Solyndra, Beacon Power, and Sage Electrochromics, LLC. One of the 13 applications we reviewed was for a project with multiple sponsors. In this instance, we only reviewed the application with the largest loan guarantee amount request.

[26] The documents LGP provided for this step indicated that the CRB decision to delegate its authority occurred on June 25, 2009. The projects in question proceeded to due diligence in May 2009.

[27] U.S. Department of Energy, Credit Policies and Procedures Manual for Implementing Title XVII of the Energy Policy Act of 2005, Revised, Loan Programs Office (Washington, D.C.: Oct. 6, 2011).

[28] 36 C.F.R. § 1222.22; U.S. Department of the Treasury, Managing Federal Receivables A Guide for Managing Loans and Administrative Debt, Financial Management Service (Washington, D.C.: 2005) and OMB Circular A-130.

[29] GAO/AIMD-00-21.3 states in part that internal control and all transactions and other significant events need to be clearly documented, and the documentation should be readily available for examination. OMB Circular A-130, Management of Federal Information Resources requires agencies to ensure that records management adequately document agency activities and ensure access to the records regardless of form or medium.

[30] U.S. Department of Energy, Office of Inspector General, Audit Report: The Department of Energy's Loan Guarantee Program for Clean Energy Technologies, DOE/IG-0849 (Washington, D.C.: Mar. 3, 2011).

End Notes for Appendix I

[1] GAO-10-627.

[2] The three excluded Mixed 2006 applications were from Solyndra, Beacon Power, and Sage Electrochromics, LLC. One of the 13 applications we reviewed was for a project with multiple sponsors. In this instance, we only reviewed the application with the largest loan guarantee amount request.

In: Federal Loan Guarantees for Energy Projects   ISBN: 978-1-62417-116-1
Editors: R. Sampson and C. Lange   © 2013 Nova Science Publishers, Inc.

Chapter 4

MARKET DYNAMICS THAT MAY HAVE CONTRIBUTED TO SOLYNDRA'S BANKRUPTCY[*]

Phillip Brown

ABSTRACT

On September 6, 2011, Solyndra, a solar system manufacturing company, filed for Chapter 11 bankruptcy protection. In September 2009, Solyndra received a loan guarantee commitment from the Department of Energy valued at $535 million, of which $527 million had reportedly been drawn down at the time of the bankruptcy announcement. Financial stress that leads to corporate bankruptcy can be caused by a number of factors, including changing market/competitive conditions, corporate and financial management decisions, financial markets, global policy changes, among others.

This report evaluates how changes in the solar market, since Solyndra's founding in 2005, might have contributed to corporate financial stress and the company's bankruptcy filing.

---

[*] This is an edited, reformatted and augmented version of Congressional Research Service, Publication No. R42058, dated October 25, 2011.

BACKGROUND

Solyndra manufactures solar photovoltaic (PV) electricity generation systems that can be installed on flat commercial rooftops. The company sells its products to value-added resellers that resell Solyndra systems to end-users such as businesses and utility companies. Unlike the majority of solar modules manufactured for electricity generation, which use flat panel silicon technology, Solyndra's proprietary PV technology is based on a cylindrical design that uses copper indium gallium diselenide (CIGS) material to convert solar energy into electricity. Solyndra claims that its unique solar system design is differentiated in the solar marketplace based on several factors that include (1) increased light collection by capturing direct, diffuse, and reflected light, (2) more electricity generation per rooftop area, (3) lower levelized cost of electricity, (4) simple, low-cost, and non-intrusive installation, and (5) reduced wind loads.

Since its founding in 2005, Solyndra has reportedly raised more than $1.5 billion: $1 billion in private investment and a $535 million loan facility, from the Federal Financing Bank, that is guaranteed by the U.S. Department of Energy. Proceeds from the DOE-guaranteed loan were used to construct Phase I of a manufacturing plant, known as Fab 2, that would be capable of manufacturing 250 megawatts per year of Solyndra solar modules.[1] Construction of Fab 2 began in September 2009 and first module shipments from the new production plant were scheduled to occur the first quarter of 2011.[2] Total cost of Fab 2 Phase I is estimated to be $733 million.[3] As of July 2011, Solyndra reportedly had sold approximately 750,000 modules throughout the world totaling roughly 100 megawatts of installed capacity.[4]

The solar PV market has experienced a number of changes since Solyndra started business operations in 2005. Around 2005, solar grade polysilicon prices began to rapidly escalate, thereby creating a strong economic value proposition for the alternative Solyndra technology. However, the market responded by adding more polysilicon production capacity and prices for the material decreased significantly.

New solar module manufacturing companies also entered the solar PV marketplace, which resulted in increased competition and price pressure for firms such as Solyndra. Solar PV module prices have declined from over $3.50 per watt in 2007 to around $1.15 per watt today. And, finally, in response to rapid solar PV price declines, European countries, Solyndra's target markets, have reduced or capped financial incentives for future solar PV projects.

In summary, the solar PV market has essentially become commoditized. Solar PV firms that have proven technology, can warranty their products, and have enough operational performance to satisfy financial risk concerns of debt and equity finance providers will be acceptable solutions for solar PV projects. Competition, therefore, will likely be based on either module prices or electricity costs. Solyndra's technology is relatively new with limited operational performance history. Furthermore, the company's announced price targets for its PV modules are higher than current market prices for competing technologies. Whether or not scaling up manufacturing would have allowed Solyndra to compete on cost is unknown. Furthermore, Solyndra's business model was narrowly focused on a niche market for which there are other possible, and potentially lower cost, solutions. With the solar PV market becoming commoditized, along with Solyndra's higher cost niche market approach, it is possible that market dynamics created a pricing environment in which Solyndra had a difficult time competing. Since Solyndra reportedly had more than $783 million of debt at the time of filing for bankruptcy,[5] the company may not have been able to profitably sell enough modules to service its debt obligations.

Following is a discussion of solar market conditions that may have contributed to Solyndra's bankruptcy filing.

POLYSILICON PRICE FLUCTUATIONS

In 2010, crystalline silicon (c-Si) solar modules made up approximately 74% of global solar module market share.[6] Solar grade polysilicon is a critical material for c-Si solar module manufacturing and an estimated 6.5 grams of polysilicon is needed per watt of wafer used to manufacture a module.[7] Polysilicon used for solar module manufacturing is typically priced in terms of dollars per kilogram ($/kg). The spot price of solar grade polysilicon experienced a high degree of fluctuation between 2003 and mid-2009 as prices ranged from as low as approximately $25/kg to as high as approximately $460/kg (see Figure 1).

The dramatic rise in solar grade polysilicon prices has been attributed to increased demand for solar PV modules in Germany and Spain starting in 2005, as a result of feed-in tariff policies and solar installation targets that incentivize installation of solar PV projects. As polysilicon prices started to escalate in 2005/2006, polysilicon suppliers responded by adding additional production capacity.

As the additional polysilicon production capacity came on line in 2008, prices started declining as supply and demand became more balanced. Since early 2009, solar grade polysilicon prices have remained relatively stable, with prices ranging from $50/kg to approximately $70/kg. On August 31, 2011, polysilicon spot prices were approximately $51.50 per kg.[8]

As indicated in Figure 1, Solyndra was founded in 2005, just as the PV market was experiencing a polysilicon shortage and polysilicon spot prices were rising.

Solyndra's technology does not require polysilicon material. Rather, the company's approach uses copper indium gallium diselenide (CIGS) material for its PV modules. By using CIGS material for its PV technology, Solyndra is insulated from solar grade polysilicon price fluctuations.

As polysilicon prices rose to as much as $460/kg in early 2008, Solyndra likely had a strong economic value proposition compared to c-Si PV modules and systems that were being challenged by high polysilicon prices.[9]

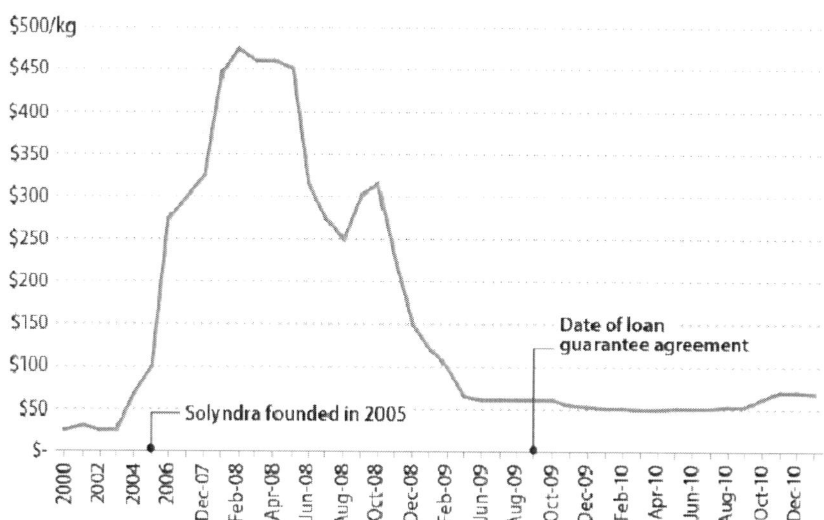

Source: Bloomberg New Energy Finance, Solyndra website, DOE loan guarantee program website.

Notes: This chart reflects historical solar "spot" prices, which may be different than "contracted" prices for solar grade polysilicon. Polysilicon suppliers may enter into long-term supply contracts for polysilicon material.

Figure 1. Solar Grade Polysilicon Historical Spot Prices.

As additional solar grade polysilicon production started coming on line in 2008, polysilicon prices began dropping and the economic value offered by Solyndra's alternative PV module approach was eroded to some degree.

Whether polysilicon price declines completely eliminated Solyndra's cost competitiveness is unknown and is beyond the scope of this analysis. In fact, it is important to note that polysilicon prices only impact the cost of c-Si PV modules and not a complete solar PV system.

The cost of an entire solar PV system includes other cost variables such as balance-of-system costs and financing costs. Neither of these variables is impacted directly by the cost of polysilicon. Nevertheless, solar grade polysilicon price declines did reduce the economic value proposition offered by the Solyndra PV solution.

ADDITIONAL PV MARKET PARTICIPANTS

Rising global demand for PV systems attracted new market participants looking to establish and solidify a position in a rapidly growing marketplace. Rising PV demand also provided an incentive for existing companies to expand production capacity.

In 2008, approximately 6.2 gigawatts of solar PV systems were installed throughout the world. In 2010, 16.6 gigawatts of solar PV systems were installed, a nearly three-fold increase in two years.[10] Most of this global market growth resulted from feed-in tariff policies in Germany, Italy, and the Czech Republic.[11]

In response to this rapid market growth, additional manufacturing capacity was added to satisfy market demand. In 2009 10.5 gigawatts of PV cells were manufactured globally and in 2010, global PV cell manufacturing capacity had reached 27 gigawatts.[12] A large portion of PV cell manufacturing capacity additions were from companies located in China and Taiwan (see Figure 2).

Additional firms in the PV marketplace may have created a much more competitive environment in which Solyndra was required to operate. The majority of the PV manufacturing capacity added in recent years has been for c-Si PV technology and the market share for this type of solar electricity generation remains above 70%.

As a result, PV system integrators and purchasers may be developing a higher degree of comfort and acceptance for c-Si technology, thus resulting in market acceptance pressure on the Solyndra technology and system.

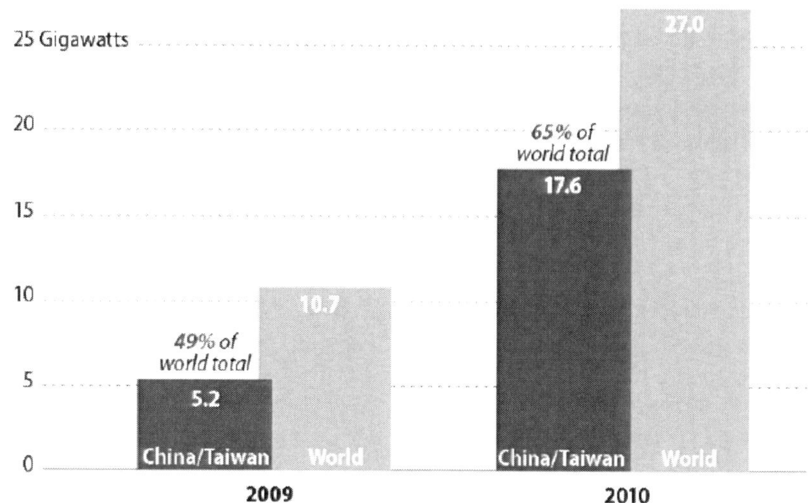

Source: REN21. 2011. *Renewables 2011 Global Status Report* (Paris: REN21 Secretariat).

Figure 2. Global Solar PV Manufacturing Capacity.

RAPIDLY DECREASING PV PRICES

As a result of decreasing solar grade polysilicon prices and additional companies entering the solar PV market, prices for PV modules have experienced steep price declines since 2007 (see Figure 3).

As indicated in Figure 3, PV module prices declined from over $3.50/watt in 2008 to approximately $1.75/watt in 2010.[13] Analysts from Greentech Media report that spot prices for solar modules were between $1.15 and $1.20 per watt in August 2011.[14] Compare this to reported statements from Solyndra management that Solyndra modules were selling for $3.24/watt in the 2009/2010 timeframe and the price target for modules was between $2.00 and $2.35 per watt.[15] Based on the forecast of global solar module prices and Solyndra's module price projections, the ability of Solyndra to profitably sustain business operations might be difficult.

Whether Solyndra's cost structure would have allowed the company to adequately respond to these forecasted market conditions, and therefore remain competitive, is unknown and beyond the analytical scope of this report. However, independent analysis of Solyndra's S-1 Securities and Exchange

Commission (SEC) filing indicates that while Solyndra was selling modules for $3.24/watt, manufacturing costs were more than $6 per watt.[16] Solyndra's negative operating margin is, arguably, somewhat common for new companies that manufacture new technologies. The business challenge for Solyndra was to scale-up manufacturing, achieve production economies of scale, and reduce costs to a level that would result in a profitable and sustainable business. However, auditor PricewaterhouseCoopers noted in Solyndra's 2010 S-1 SEC filing that "the company has suffered recurring losses from operations, negative cash flows since inception and has a net stockholders' deficit that, among other factors, raise substantial doubt about its ability to continue as a going concern."[17]

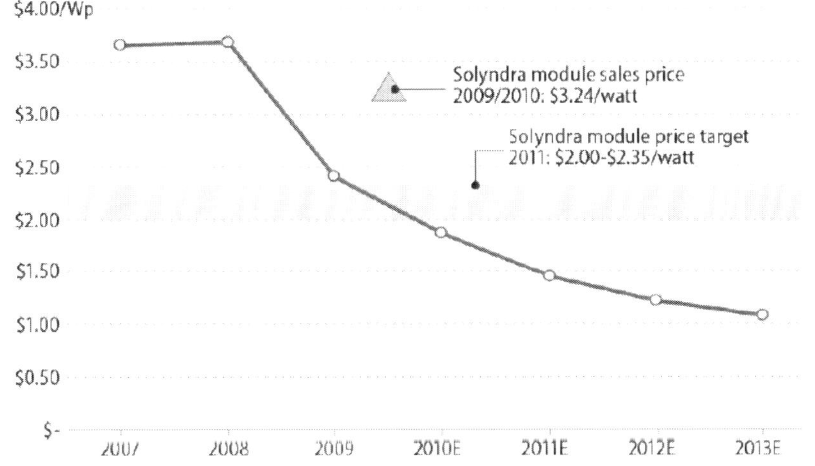

Source: Greentech Media.
Notes: Wp = watt peak. Blended average selling prices include prices for all types of solar module technologies: c-Si, cadmium telluride, CIGS, amorphous silicon, etc. E = estimated.

Figure 3. Global Blended PV Module Average Selling Price; (2007-2013E).

COMMERCIAL ROOFTOP NICHE MARKET

Solyndra's business model was solely focused on addressing potential demand for solar electricity generation on commercial building rooftops. The Solyndra technology design provided this niche market with some unique characteristics: more rooftop area covered with solar modules, reduced wind loads, and simple, non-intrusive, and low cost installation, among others.

According to Solyndra's 2010 S-1 SEC filing, approximately 11 billion square meters of commercial building rooftop area exists throughout the world.[18] This rooftop area represents, perhaps, the entire theoretical market potential for Solyndra systems. However, what is not clear is how much of this rooftop area is an addressable market. Many commercial buildings are located in areas that have poor solar resources and, therefore, would not justify investment in a rooftop solar PV system. Furthermore, economic, policy, and other market conditions may eliminate some portion of commercial building rooftops from the addressable market. Exactly how much global rooftop area is actually addressable is not known at this time.

To date, many solar PV installations have been on residential rooftops and a growing solar PV market segment is classified as "utility scale." While there is no formal definition of this market segment, utility scale solar PV projects might be 10 megawatts or larger and consist of ground-mounted PV systems. The Solyndra technology is not ideally suited for residential or utility scale types of projects and, as a result, the company was limited in its options to diversify into other solar PV markets if necessary.

Finally, Solyndra's technology is not the only solution available for commercial rooftop PV projects. Flat-panel silicon and thin-film technologies can also be installed on commercial rooftops. Solyndra was positioning its technology as one that can provide the lowest levelized cost of electricity, as its configuration allows for more rooftop area to be covered with electricity generating photovoltaics. However, the current operational efficiency of Solyndra's CIGS technology is approximately 11% to 12% compared to c-Si efficiencies of approximately 14.3%, which is 20% higher relative to CIGS.[19] Therefore, for a given commercial rooftop, higher efficiency silicon PV could potentially generate as much electricity as the Solyndra CIGS approach even though Solyndra's technology would cover a larger portion of the commercial rooftop area. This efficiency differential combined with consistent silicon PV cost and price reductions may have resulted in Solyndra losing its position within the commercial rooftop niche market.

INCENTIVE DECLINES IN EUROPEAN MARKETS

According to Emerging Energy Research, Solyndra was emphasizing European commercial rooftop markets. This is evidenced by framework supply agreements with six European developers with a potential value of $1.57 billion, and that represents 71% of its announced supply agreements as of December 2009.[20]

European solar markets are typically incentivized through feed-in tariff policies that essentially provide guaranteed rates for electricity generated from solar or other renewable technologies over a period of 10 to 25 years.[21] Feed-in tariff rates and qualification criteria differ from country to country. However, feed-in tariffs are typically generous enough to provide project developers with investment rates of return high enough to incentivize solar project installations. The most notable European countries with feed-in tariff incentives include Germany, Spain, France, Italy, and Czech Republic. Table 1 summarizes the amount of solar PV capacity added in these countries, and the U.S., which does not have a federal feed-in tariff policy, since 2006.

Table 1. Solar PV Additions: 2006-2010
(in megawatts)

	2006	2007	2008	2009	2010
Germany	845	1,270	1,950	3,795	7,405
Spain	90	560	2,600	145	370
Italy	10	70	340	715	2,320
Czech Republic	—	3	60	400	1,490
France	10	10	45	220	720
United States	145	205	340	475	880

Source: REN21. 2011. *Renewables 2011 Global Status Report* (Paris: REN21 Secretariat).

As indicated in Table 1, Spain experienced a solar market boom and bust between 2008 and 2009. The large amount of solar PV additions in 2008 can be attributed to a feed-in tariff policy that was extremely generous and did not have caps or other mechanisms to control explosive growth. As a result, project developers rushed into the Spanish market to take advantage of lucrative incentives. However, in 2008 the Spanish government revised payment levels down and capped its feed-in tariff incentive policy. The market for solar PV in Spain declined dramatically in 2009.[22]

Germany has experienced consistent solar market growth since 2006 and had record solar PV additions of 7,405 megawatts in 2010, nearly double the additions in 2009. Germany's feed-in tariff policy incorporates responsive degression, which includes a base feed-in tariff reduction every year as well as additional adjustments, either up or down, based on the amount of capacity installed each year.[23] Germany's degression approach is structured to be responsive to the rapid price and performance changes occurring in the global solar marketplace. As a result of large solar additions during 2010, Germany's feed-in tariff will be further reduced, based on the defined degression schedule, therefore reducing the rate-of-return premiums that might otherwise be available. Germany's feed-in tariff reductions also result in a challenging economic environment for higher priced, niche market products such as those sold by Solyndra.

Incentive changes and modifications in European countries, one of Solyndra's target markets, may have limited the company's ability to sell products into the respective markets.

End Notes

[1] Solyndra S-1 SEC filing, available at http://www.sec.gov/Archives/edgar/data/1443115/000119312510058567/ ds1a.htm#toc15203_8.

[2] Ibid.

[3] Solyndra S-1 SEC filing, available at http://www.sec.gov/Archives/edgar/data/1443115/000119312510058567/ ds1a.htm#toc15203_8.

[4] Kanellos, Michael, "Stat of the Week: 0.2 percent," Greentech Media, July 15, 2011.

[5] "Solyndra files for bankruptcy, looks for buyer," Associated Press, September 6, 2011.

[6] Mehta, Shyam, "PV Technology, Production and Cost Outlook: 2010 –2015," Greentech Media, January 5, 2011.

[7] Kim, Anthony, "Analyst Reaction – Solar Price Surveys, January 2011," Bloomberg New Energy Finance, January 17, 2011.

[8] http://pvinsights.com/.

[9] According to Bloomberg New Energy Finance, approximately 6.5 grams of polysilicon is needed per watt of solar wafer. At $460/kg, polysilicon raw material would cost nearly $3.00 on a per-watt basis. At $50/kg, polysilicon raw material would cost approximately $0.33 on a per-watt basis.

[10] REN21. 2011. *Renewables 2011 Global Status Report* (Paris: REN21 Secretariat).

[11] Ibid.

[12] Ibid. It is important to note that cells and modules are different products. PV cells are typically assembled, electrically connected, and integrated onto a substrate material to form a solar module. Module manufacturing capacity may be different than PV cell manufacturing capacity.

[13] Mehta, Shyam, "PV Technology, Production and Cost Outlook: 2010 –2015," Greentech Media, January 5, 2011.

[14] Kanellos, Michael, "Will Solyndra, or Part of It, Get Bought?," Greentech Media, August 31, 2011.

[15] Kanellos, Michael, "Solyndra to Dop by 50% in Price by 2012, Says CEO," Greentech Media, November 3, 2010.

[16] Mehta, Shyam, "Solyndra: 1.9 MW Project Installed, But Story Remains Fraught With Uncertainty," Greentech Media, February 1, 2010.

[17] Solyndra S-1 SEC filing, available at http://www.sec.gov/Archives/edgar/data/1443115/00011 9312510058567/ ds1a.htm#toc15203_8.

[18] Ibid.

[19] Mehta, Shyam, "PV Technology, Production and Cost Outlook: 2010 –2015," Greentech Media, January 5, 2011.

[20] T. Maslin and C. Deline, "US-Based Solyndra Specializes on EU Rooftops," Emerging Energy Research, December 30, 2009.

[21] For more information regarding feed-in tariffs see T. Couture, K. Cory, C. Kreyik, and E. Williams, "A Policymaker's Guide to Feed-in Tariff Policy Design," National Renewable Energy Laboratory, July 2010.

[22] Karlynn Cory, "Stops and Starts in the Spanish Solar Market," National Renewable Energy Laboratory, November 30, 2009, available at http://financere.nrel.gov/finance/content/ stops-and-starts-spanish-solar-market.

[23] For more information regarding Germany's solar feed-in tariff degression approach, see "EEG Amendment 2012 – What will change as of 1 January 2012," BSW Solar, available at http://en.solarwirtschaft.de/fileadmin/content_files/ EEG-Novelle2012_EN.pdf.

INDEX

D

E

Y